Higher Education and Silicon Valley

HIGHER EDUCATION

AND

SILICON VALLEY

Connected but Conflicted

W. RICHARD SCOTT,
MICHAEL W. KIRST,
AND COLLEAGUES

Johns Hopkins University Press

BALTIMORE

© 2017 Johns Hopkins University Press
All rights reserved. Published 2017
Printed in the United States of America on acid-free paper
9 8 7 6 5 4 3 2 1

Johns Hopkins University Press
2715 North Charles Street
Baltimore, Maryland 21218-4363
www.press.jhu.edu

Library of Congress Cataloging-in-Publication Data
Names: Scott, W. Richard, author. | Kirst, Michael W., author.
Title: Higher education and Silicon Valley : connected but conflicted /
    W. Richard Scott, Michael W. Kirst, and Colleagues.
Description: Baltimore : Johns Hopkins University Press, 2017. |
    Includes bibliographical references and index.
Identifiers: LCCN 2016046943| ISBN 9781421423081 (pbk. : alk.
    paper) | ISBN 1421423081 (pbk. : alk. paper) |
    ISBN 9781421423098 (electronic) | ISBN 142142309X (electronic)
Subjects: LCSH: High technology and education—California—San
    Francisco Bay Area. | High technology industries—Employees—
    Training of—California—San Francisco Bay Area. | Business and
    education—California—San Francisco Bay Area. | Education,
    Higher—Economic aspects—California—San Francisco
    Bay Area.
Classification: LCC LC1087.3.C2 S36 2017 | DDC
    378.1/035097946—dc23
LC record available at https://lccn.loc.gov/2016046943

A catalog record for this book is available from the British Library.

*Special discounts are available for bulk purchases of this book. For more
information, please contact Special Sales at 410-516-6936 or specialsales
@press.jhu.edu.*

Johns Hopkins University Press uses environmentally friendly book
materials, including recycled text paper that is composed of at least
30 percent post-consumer waste, whenever possible.

Contents

Preface

The eyes of the world are on the San Francisco Bay Area. Much of the attention is focused on the region's economy, trying to understand and replicate a remarkable success story in the information technology sector. The world, however, is also closely watching the Bay Area's higher education field. Hewlett-Packard, Cisco, Apple, and Google are not the same types of manufacturing or service giants that dominated previous economic booms in the United States; their dynamics are driven by intellectual capital, which requires high levels of skills and training.

While the technology sector and higher education are similar in some ways and interdependent, they are also mismatched in several important respects. The two fields developed under differing conditions, have been responsive to different constraints and incentives, and differ substantially in their pace of change. The institutions of higher education in this country developed in the latter part of the nineteenth century and adopted a bureaucratic structure that provided discretion to academic professionals. The ends served are intangible and remote. In contrast, the structure of new industries is made up of multiple, flexible organizations adapted to respond rapidly to changes in market opportunities and demands. To understand the interactions between the Bay Area's higher education providers and the Silicon Valley economy, it is essential to view these organizations within the context of their differing fields.

Unlike many other accounts linking higher education and hi-tech industries, our approach is not restricted to the role of the research universities such as Stanford and the University of California, Berkeley, although they are of obvious importance. Rather, we consider the full range of academic institutions, focusing especially on "broad-access" colleges, like the comprehensive state colleges, nonprofit and for-profit colleges, and community colleges that enroll the great bulk of students.

By incorporating broad-access colleges in the conversation, we have found a more complex ecology that holds implications for regions that lack monolithic ivory towers. Broad-access colleges may not create breakthrough technologies, but they do provide vital human resources to sustain the development of industry. These institutions have been frequently overenrolled and underfunded, and subjected to financial and administrative constraints that prevent them from fully responding to the needs of regional and national economies. They warrant attention now more than ever.

Our study comes in the context of increased pressure on colleges and universities to offer more programs featuring career and vocational training. Institutions are responding both to the expressed desires of their clients and to a clamor from national, state, and regional interests demanding that higher education do more to stimulate economic development and provide workforce training. This impetus is not new, but it has become more intense in recent decades. More importantly, because of the huge variety among types of colleges and variation in local conditions, there exists increasing variability both in the kind of economic and political pressures being exerted and in the strategies pursued by colleges in responding to them.

It is impossible to understand developments occurring within higher education in our time without attending to changes in the ecology of the organizations providing these services. The nature and types of colleges are changing, as is the constellation of the support and control systems they confront. To accommodate this variety, this book takes a longitudinal approach, integrating data from interviews, statistical analysis, and document analysis that cover more than 45 years, beginning in 1970. Our aim is to chronicle changes in the types of colleges operating in the Bay Area during this time frame, as well as changes over time in the numbers and types of hi-tech employers. We also examine internal changes in colleges occurring during this period in the types of faculty employed, the academic programs offered, and the types of graduates who emerged. Throughout, we endeavor to address the effects of the larger historical context, including policy decisions at federal and state levels, administrative structures, and reactions from faculty and students, on the events and trends that we describe.

Additionally, although the Bay Area is geographically compact, we found addressing it as a single entity was misleading. At times throughout the book, we break down the larger Bay Area into three sub-regions, each of which has its own unique ecology of industry and higher education, as well as its own unique demographic patterns. This sub-regional approach helps us to explore in more detail *who*—what types of colleges—adapted in an attempt to meet the changing needs of local employers—and *how*—what strategies and tactics colleges employed to adapt to industry demands. We also take care to include different institutional forms in our examination of sub-regions, paying close attention to public institutions (both two-year and four-year), private nonprofits, and for-profits in each area.

The end result, we hope, is a book that is relevant for anyone engaged in the broad enterprise of higher education including administrators, educators, academic researchers, and students, as well as those working in professional associations and public and private oversight bodies. Our intended audience includes business leaders, human resource staff, planners, and economic geographers interested in understanding and improving the interface between schools, training, and work. Students and scholars in the areas of higher education, organizational sociology, and regional economics will find much of interest here. Finally, policy makers at all levels may find this book a stimulating repository of lessons about the ways in which higher education institutions and industry can collaborate—or why they fail to do so.

As an exemplar of hi-tech regional development, the San Francisco Bay Area commands wide interest, but some may wonder if this study holds interest to those living and working in other areas with differing industrial/economic structures. We believe that it can serve such groups by providing a template for study. The conceptual approaches embraced and the longitudinal methodologies employed can be readily transferred to the examination of diverse regional economic systems. Industrial systems and demands and the composition of educational providers will differ from region to region, but regardless of their make-up it will be worthwhile to examine how the economic and educational fields confront and adapt to one another.

Regional approaches are especially important when considering "local industries" like higher education. Although major research and comprehensive universities draw students from far outside regional boundaries, the great majority of colleges serve students in nearby areas. Hillman (2012) reports that the typical student in a community college traveled eight miles; older students returning to college or seeking technical training or retraining also rely principally on the programs available in their immediate vicinity. Hence, we would expect the numbers and the types of colleges and other postsecondary schools and the courses they offer to vary substantially by the specific mix of industries in their regions.

The introductory chapter that follows this preface details our research questions, study design, and methodology, finishing with a preview of the policy implications of our research. Chapter two describes the higher education sector as it appears in the Bay Area through an organizational field perspective, focusing both on types of colleges and on the many extramural organizations and forces that both support and control their operations. Chapter three also employs a field perspective but shifts the focus to industry, describing the regional economy of the Bay Area, again through an organizational field lens, and with special attention to the emergence and transformation of the "Silicon Valley" labor market. The fourth chapter examines the broader forces at the state and national levels that shape both higher education and the regional economy, including demographic shifts, lifestyle changes, technological and organizational developments, and political and economic forces. Chapter five drills down to example three sub-regions within the Bay Area and presents portraits of selected colleges and universities in each area. Chapter six takes a close look at the different ways in which the colleges and universities in our study have adapted to changes in the ecology that surrounds them. Finally, we end with a concluding chapter that identifies the major issues confronting higher education, describing implications for public policy and suggesting some new initiatives, at national, state, and regional levels, which could alleviate some of the friction and conflict identified in our study. Throughout the book, we illustrate our narrative with in-depth case examples of phenomena and institutions that shed light on our argument.

While we sometimes take a normative tone, this book is meant to be the beginning of a conversation, not the final word. We envision that conversation as one that takes a broad view of the relationship between higher education and the economy, not one rooted exclusively in the internal functions of colleges and universities, or in the demands of employers. Those who manage, enact, study, and care about the future of higher education have much work to do, and we hope that the chapters that follow provide an expansive and integrative framework to guide the debates and initiatives to come.

This study is the work of multiple authors, but still does not represent all of those who have contributed to the completion of this enterprise. Our study was grounded in an early project, "Reform and Innovation in the Changing

Ecology of U.S. Higher Education," under the leadership of Michael W. Kirst and Mitchell L. Stevens, funded by the Bill & Melinda Gates Foundation. W. Richard (Dick) Scott and Thomas Ehrlich were senior advisors for this effort, and Kristopher Proctor, Rachel Baker, and Daniel Klasik were important contributors to the intellectual development of this work, which was designed to enlist the aid of creative scholars to initiate a wider discussion and create a research framework for higher education in the twenty-first century (Kirst and Stevens 2015).

With grant support from the S. D. Bechtel Jr. Foundation, the current project commenced in 2013 under the direction of principal investigator Mike Kirst and senior scholar Dick Scott. Our research received additional support from LearningWorks. Team members from the John W. Gardner Center for Youth and Their Communities at Stanford University's Graduate School of Education were instrumental in the management and implementation of the research. Rebecca London served as the manager of the project at its inception, but was succeeded by Manuelito Biag early in our work. Laurel Sipes and Judy Liang helped review the literature and policy landscape, as well as collect and analyze data. Nancy Mancini provided strategic support regarding communications.

In addition to Gardner Center staff and our coauthors—Manuelito Biag, Brian Holzman, Bernardo Lara, Anne Podolsky, and Ethan Ris—other researchers included Elizabeth Dayton, Steven Gentile, and Carolina Ornelas. Mitchell L. Stevens and Thomas Ehrlich provided useful feedback and sound guidance throughout the course of our work, and Francisco Ramirez and Hans N. Weiler offered valuable insights on an earlier manuscript. As described in chapter one, our project also benefitted from the experience and wisdom of a number of seasoned educational leaders and scholars in the San Francisco Bay Area who provided orientation during the shaping phases of our research. These individuals are listed in appendix A.

Higher Education and Silicon Valley

Introduction

W. RICHARD SCOTT, MICHAEL W. KIRST, MANUELITO BIAG, AND LAUREL SIPES

Jennifer boards her company's shuttle bus at the stop a block from her apartment in San Francisco. Coffee in hand, she seeks a window seat. Unlike many of her colleagues, she prefers not to work or play games on her phone during the ride. She moved to the Bay Area from Taiwan six months ago and still is struck frequently by the beautiful landscape. As the bus crawls through the typical southbound commuter traffic on Highway 101, Jennifer sees the sign for City College of San Francisco. Yesterday she was looking at the spring course offerings, considering enrolling in the Ruby programming language course. It was certainly less expensive than the coding bootcamps that some of her friends were doing, though she does like the digital badges they can earn and include in job applications. She wonders if it would be difficult to make it to campus one evening a week.

As the bus continues its slow progress down the peninsula, Jennifer notices one school after another that would help students meet all sorts of educational goals. The A-1 Trucking School in Burlingame, Marinello School of Beauty in San Mateo, Notre Dame de Namur University, and the University of California Berkeley Extension in Belmont. In Foster City, she sees that the University of Phoenix has a campus. As the bus passes Palo Alto, she knows that Stanford University stands behind the eucalyptus trees lining the highway. She is surprised to see that Carnegie Mellon has a Silicon Valley campus in Mountain View. More familiar to her along the route are the many big-name technology firms that are regularly discussed as drivers of economic growth and innovation: Twitter and Dropbox in San Francisco, Genentech in South San Francisco, YouTube in San Bruno, Oracle in Redwood City, and Facebook in Menlo Park. She has new friends and acquaintances who work at most of them. As they grow closer to her company's campus, Jennifer notices even more schools: Mission College, Santa Clara University, California College of Nursing, the Art Institute of California, and San José State University. The bus also passes even more big-name firms: Lockheed Martin, Google, Yahoo, Dell, and Cisco.

At the end of the too-long ride, Jennifer finally arrives at her desk. She thinks about the students who attend all those schools, and the workers at all those technology firms. She wonders how many students move from a local college or university to jobs in the Bay Area tech sector. She wonders how many of her colleagues did. Jennifer was recruited from Taiwan and works in the United States on an H-1B visa. She was surprised and pleased when she arrived to find an Innovation and Entrepreneurship Center for Taiwanese workers in Silicon Valley. She appreciates their networking events and the advanced English course that she took as soon as she arrived. If the mix of people in her company's dining space at lunch is any indication, her coworkers hail from all over the world. But some, she muses, must come from right here.

After checking her Twitter, Instagram, and Facebook feeds, Jennifer settles in to her morning tasks. As she ticks through her new email, she notices that she has a message from LinkedIn. They ask whether she would be willing to act as a virtual mentor to students who attend her alma mater and are interested in pursuing a similar career. Jennifer would like to help. She wonders if she has the time on top of that Ruby course.

―――――――――――――――――

New and ascending economic and political forces are challenging the field of higher education in America. Previously dominant academic and professionally governed forms are increasingly subjected to economic and political pressures that urge the adoption of different criteria for generating curricula, hiring faculty, and evaluating students. In short, a major sorting out is currently underway concerning what criteria—what "institutional logics"—are to prevail in designing colleges. We investigate this contest using an approach that differs in several respects from previous research.

Most studies of higher education have concentrated attention on highly visible research universities or selective baccalaureate colleges, ignoring the vast number of "broad-access" institutions that serve the great majority of students in US higher education (Stevens 2015). In addition, even when scholars examine a wider range of institutions, most concentrate on changes occurring over relatively short periods of time. Finally, very few studies examine higher education programs as they relate to their immediate economic context. In contrast, we propose an inclusive framework that incorporates the full range of higher education institutions operating in one metropolitan region over several decades. Our study concerns the changes that have occurred in the field of higher education in the diverse and dynamic San Francisco Bay Area from the 1970s to today.

Higher Education in the San Francisco Bay Area

The greatest long-term threats to Silicon Valley are . . . continued reductions in public funding for California's educational institutions—from its elementary and secondary schools to its sophisticated network of community colleges, state universities, and the University of California [UC] system. [They] jeopardize the rich supply of technical talent and the research base that have historically supported the regional economy.   (SAXENIAN 1996: IX)

Colleges and universities are increasingly seen as important players in the economies of the regions where they are located. In turn, regions have recently come to be viewed as critically important for supporting widely varying economic activities (Moretti 2012). These dynamics are nowhere more apparent than in the San Francisco Bay Area, the home of Silicon Valley. As the seat of a world-famous innovation economy, this region's success rests upon the knowledge and training of its workforce. This would suggest that postsecondary institutions in the area ought to be highly valued as vital resources to the region's economic growth. Many observers have called attention to the contributions of research universities, such as Stanford University and UC Berkeley, in the development of the region's economy over the past 50 years. Here, we seek to understand the roles *all* types of postsecondary institutions in the Bay Area have played and continue to play in developing and sustaining the technology sector in Silicon Valley and the broader San Francisco region.

Adaptation by the various colleges to the demands of a dynamic economy has been made more difficult by the twin pressures of expanding enrollment demands and reduced resources. California has experienced rapid population growth, expanding more than 250 percent since 1950, and the Bay Area has grown even more rapidly during the past four decades. State support for public colleges has failed to keep pace with rising demands for higher education. The gap between the number of Californians applying for admission to four-year public colleges—both research and state universities—and those admitted has more than doubled since 1996 (Campaign for College Opportunity 2015).

Silicon Valley has received worldwide attention as one of the primary—if not the premier—examples of a vibrant, entrepreneurial, and innovative region (Kenney 2000; Lee et al. 2000; Malone 2002; Saxenian 1996). Since the 1960s, Silicon Valley has undergone numerous reinventions, beginning with defense and aerospace industrial development, followed by semiconductors, personal computing, the Internet, social media, and biotechnology (Henton 2000; Henton, Kaiser, and Held 2015). Several distinctive features characterize this rapidly changing economy, including flexible labor markets that incentivize workers to constantly upgrade their skills and to move from one firm to another; a mix of small, medium, and large companies; and a reliance on venture capital financing (Barley and Kunda 2004; Benner 2002).

Until recently, most colleges and universities have given higher priority to academic subjects and training than to practical, applied subjects and real-world experience. They have stressed mastery of knowledge as more important than acquiring sets of skills, although, as we will emphasize, important differences in the degree of this prioritization exist among the various types of colleges and individual schools in the region. Increasingly, postsecondary institutions are experiencing pressure to orient more of their energies to preparing their graduates to participate effectively in the workforce and contribute to the development of the economy (Brint 2002; Kirst and Stevens 2015). This pressure represents a shift away from the traditional mission of most American colleges and universities. This newer "demand side" view emphasizes that the buyers (in this context, employers) are the ones who should determine the value of skills and knowledge of a successful graduate.

The traditional "supply side" model stressed the importance of providing postsecondary graduates with a liberal education so that, as informed citizens, they could contribute to the well-being of society. In this model, the providers (educators) seek to uphold internal academic norms and practices and to compete with one another for academic prestige and standing, thereby determining what students learn and how they learn it. In many respects, this model is a product of forces and processes at work over the past two centuries. Templates for American educational systems were crafted during the period of industrialization and modernization in the eighteenth and nineteenth centuries, which gave rise in the twentieth century to large, stable, bureaucratic organizations. Although incorporating some craft and professional components, such as greater autonomy for faculty and elements of collegial control, most systems of higher education are characterized by elaborate hierarchies, extensive rule systems, and settled routines of work (Blau 1973; Clark 1983). Higher education systems value tradition and continuity, ritual and ceremony, and are prone to take long-term views of their mission and goals. The ability of colleges to adapt to changing demands is also affected by declining support from public funding, particularly at the state level. In California, funding for state colleges has declined in recent decades, and, particularly in the Bay Area, it has failed to keep pace with demographic changes and increasing demand, making it difficult for them to adequately respond to changing conditions and expectations.

In contrast, technology firms within Silicon Valley are increasingly looking for employees with demonstrable, relevant skills, in addition to traditional credentials (Carnevale and Desrochers 2001; Casner-Lotto and Barrington 2006; Gallivan et al. 2004). They prize specialized up-to-date know-how and the latest skills over diffuse, general knowledge. They value innovation and dynamism, and are engaged in pursuing the next new thing. Given the dominant features of these two arenas—higher education and the technology economy—we observe a substantial tension in their values, norms, practices, and the pace of change. As the president of one Bay Area state university acknowledged, "the regional economy changes exponentially, but my university can only change incrementally." And, as noted, such adaptations are difficult because of inadequate funding at the state level.

The contrast in values is not absolute. Many company managers insist that they recognize the value of colleges beyond simply providing a source of technical skills. They understand and recognize the importance of clarity of thought, the ability to write and speak clearly and forcefully, the capability to learn, and the capacity to collaborate—all abilities associated with a liberal education (Zakaria 2015). Many firms recognize their need for literate managers and leaders at all levels, not simply technicians, programmers, and accountants. Nevertheless, the demands for a highly skilled technical workforce loom large.

This study examines the types of cross-pressures operating on postsecondary educational institutions attempting to both uphold academic standards and honor the traditions of the past on the one hand, while striving to respond to the demands of a volatile and rapidly changing market economy on the other. It also recognizes the great variability among types of colleges in their missions, structures, and capacity to adapt to market demands. In addition, we detail some of the kinds of strategies pursued by specific colleges as they respond to their changing context.

Applying an Organization Field Lens

Most studies of higher education choose the individual student as their primary focus. Sociologists have long regarded higher education as the fundamental motor of individual mobility in modern societies (Sewell and Hauser 1975). Economists typically view education as the primary way human capital is created and economic development advances (Becker 1964; Freeman 1982). Hence, the lion's share of relevant social science research as well as coverage in the popular media is devoted to examining how and to what extent individuals and particular subpopulations benefit from years of schooling. Our study, in contrast, focuses attention on the colleges themselves: the entities that employ the faculty, select the academic programs, admit students, conduct the instruction, and determine who graduates.

Even when the focus is on colleges, until recently most researchers have focused on "elite" schools—highly selective liberal arts colleges such as Oberlin or Antioch (e.g., Clark, 1970; Karabel 2005) or on research universities such as Columbia or Stanford (e.g., Cole 2010; Lowen 1997)—even though these types of schools make up less than 15 percent of higher education programs and educate fewer than 10 percent of college students in the United States. While our study does not neglect the important role played by these colleges and universities, we concentrate attention primarily on *broad-access colleges*—those schools that admit the majority of their applicants. These public and private programs, varying greatly in size, mission, and structure, educate the vast majority of postsecondary students in this country. Indeed, over half of all US college students enrolled today attend two-year community colleges, a type of school that barely existed 50 years ago. By focusing more attention on broad-access colleges, our work contributes to the emergence of a small but growing number of studies of these institutions at the organizational level (e.g., Arum and Roksa 2011; Bailey, Badway and Gumport 2002; Rosenbaum, Deil-Amen, and Person 2006).

In our study, we adopt an *organization field* perspective. This approach reminds us that the most important aspect of the environment in which organizations operate is other organizations. Since any segment of modern society contains many organizations (March and Simon 1958), each organization must decide which subset is most salient for a given issue. To pursue greater understanding of the tensions already described, we find it helpful to think of every college as operating in at least two basic organizational subsystems:

1. A sectoral slice that highlights those organizations that are most similar (other colleges) along with those related organizations exercising support and/or control functions. The field of higher education contains a variety of types of colleges and a wide range of professional and regulative systems that encourage colleges to conform to academic values and traditions but, at the same time, engender competitive forces that press colleges to act strategically in attracting faculty and students.

2. A regional focus that emphasizes the interdependence of dissimilar organizations that operate in the same locality. All educational programs necessarily participate in the local economy and, of particular interest for our study, develop relations with those organizations that utilize their outputs (graduates and knowledge). The field of a regional economy is made up of suppliers and buyers of human capital as well as a range of supporting organizations, such as placement service agencies, serving primarily as intermediaries or brokers. This system is mainly governed by market mechanisms, although Silicon Valley is recognized as hosting multiple types of interorganizational and interpersonal networks. There have been also modest attempts at regional coordination and planning.

The two fields of interest thus incorporate both the supply-side, dominated by professional academic interests, and the demand-side, driven by market pressures, allowing us to juxtapose them and examine their incompatibilities, complementarities, and intersections.

All fields entail both *isomorphic* pressures, which constrain organizations to adopt structures and processes similar to their counterparts, and *competitive* pressures, which reward organizations for behaving in less conforming, more strategic ways. While a sectoral conception of fields puts primary emphasis on isomorphic pressures conducing to conformity, it also accommodates competitive processes. Regional conceptions, which necessarily incorporate a wider range of types of organizations, more readily recognize competitive forces. Early organization field conceptions, such as Meyer and Rowan (1977) and DiMaggio and Powell (1983), accorded too much emphasis to isomorphic pressures, assuming that organizations would readily conform to dominant field norms. Newer versions, which we espouse, note that actors, both individual and organizational, are both constrained and empowered by institutional frameworks—prevailing rules, norms, and cultural beliefs—and that they are capable of using them to pursue their own interests as well as to challenge and attempt to change institutional frameworks if necessary (Scott 2014). Fields include both established players who have a vested interest in maintaining the status quo, as well as challengers whose interests have been suppressed but seek ways to mobilize to promote change and reform. Far from being islands of tranquility and harmony, fields are contested arenas in which actors with diverse interests and agendas struggle for resources and influence (Fligstein and McAdam 2012).

Colleges within a region differ in their structure and mission and so experience these field forces in varying ways. Fields consist of more than the numbers and types of social actors (individuals and organizations) and the types of relations among them. They are also cultural systems in which actors hold varying values and respond to varying norms: they embrace differing *institutional logics*. Institutional logics are the shared conceptual and normative frameworks that provide guidelines for the behavior of field participants (Friedland and Alford 1991; Thornton and Ocasio 2008). Liberal arts colleges and academic departments within comprehensive colleges and research universities have long upheld the ideal of a liberal education in which learning is viewed as an end in itself, as increasing one's knowledge of the world and its variety, developing judgment and refined sensibilities, and as preparation for responsible civic participation and leadership. This institutional logic has guided the design of academic departments, content of course curricula, and standards for degree conferral.

Specialized institutions, including vocational and professional programs, have mostly operated at the margins of the higher education field. Clearly, these programs are more attuned to the technical and market logics of industry. But during recent decades we have seen the rise of community colleges, which are designed to be hybrid institutions, providing liberal arts programs, which enable students to transfer to four-year colleges, and also providing technical programs, permitting students to obtain terminal two-year programs and certificates in vocational areas and adult education supporting retraining. Similarly, comprehensive colleges, such as state universities, have added large numbers of professional schools and technical programs targeting vocational training. These types of institutions endeavor to simultaneously serve both logics: liberal arts education and practical training.

In short, there exist a wide variety of postsecondary programs, and they are scattered along a continuum ranging from those emphasizing basic liberal and classical education at one end to those specializing in vocational and workforce training at the other. Increasingly, most colleges host some mixture of these two types of curricula in their portfolio, although, over time, this mixture is gradually shifting to favor the vocational side.

Studying the San Francisco Bay Area

The San Francisco Bay Area is host to Silicon Valley, a globally prominent region of technological innovation and development that emerged soon after the Second World War. Giants of the technology industry, such as Hewlett-Packard, Fairchild Semiconductor, National Semiconductor, Lockheed Martin, Apple, Cisco Systems, Intel, Yahoo, Google, Facebook, and many more, were founded and operate here. In addition, a number of venture capital organizations have arisen to provide important sources of revenue to emergent firms. Other types of organizations that play an important role in the innovation economy include federal laboratories such as Lawrence Livermore and NASA's Ames Research Center, independent and corporate labs such as those operated by SRI and PARC, and a range of intermediaries such as temporary employment agencies, nonprofit employment training and placement services, business advocacy organizations, and professional associations (Benner, Leete, and Pastor 2007; Randolph 2012).

Colleges and universities of all types have played a variety of roles in the Silicon Valley economy during this period. Stanford and the UC Berkeley have been vital sources of ideas and human capital, both technical and administrative. Even more important, many faculty members and graduates have played critical roles, not only as technical innovators but also as entrepreneurs, collaborating with others to launch new firms. In 1996, start-ups involving Stanford faculty accounted for nearly 60 percent of total Valley companies (Gibbons 2000; Lenoir et al. 2004). Broad-access colleges have also played crucial roles. For example, according to its president, San José State University graduates between 600 and 700 engineers each year, and most are employed in the Valley. This type of economy requires a range of qualified workers, not simply innovative engineers and entrepreneurial leaders but also software programmers, tech-savvy marketing personnel, accountants, various types of managers, and other workers with midlevel skills. This perspective on the demands of the technology industries in the Valley and the broader Bay Area encourages us to consider the competitive forces that induce individual colleges serving a range of students to act in strategic ways to respond to the needs of this economy. Colleges and their programs are in the process of adapting to their regions.

Research Questions and Study Design
Guiding Research Questions

- How has the organizational ecology of higher education in the San Francisco Bay Area changed from 1970 to the present? How have the numbers and types of postsecondary institutions in the Bay Area and the relations among them changed during this period?
- How has the internal structure, types of students served, and programs of Bay Area colleges changed during this time?
- How has the regional economy of the Bay Area changed during this period—in particular, that part of the economy associated with hi-tech industries?
- How have the wider state and federal systems affecting higher education changed during this period, and to what extent have Bay Area colleges adapted to these broader demographic, economic, and political transformations?
- How and to what extent did colleges and other postsecondary institutions become engaged in the Bay Area's hi-tech regional economy from 1970 to the present day? What contributions have these colleges made to meeting the labor force needs of this region?
- To what extent were postsecondary institutions and Bay Area industries linked? Through what mechanisms did these connections occur? What strategies did each sector use to ensure adequate preparation of students for future career pathways?
- How have the national, state, and regional policies and programs supporting and guiding the operation of higher education changed during this period, and what are the questions and concerns that need to be addressed in moving forward?

Research Design

The broad design for the study has three phases.

Phase 1: Exploratory Interviews with Expert Informants

To inform ourselves about the current nature of higher education in the Bay Area and related environmental developments, we conducted exploratory interviews with a variety of expert informants, including those in leadership positions in various types of colleges and their professional schools, experts on the Silicon Valley economy, and administrators from local postsecondary institutions (see appendix A). These interviews helped us refine our research questions and approach prior to finalizing our research design and data collection instruments.

Phase 2: Quantitative, Longitudinal Study of Higher Education and the Regional Economy in the Bay Area

To document the complexity and constantly changing ecology of postsecondary institutions during our more than four-decade period of study, we gathered data from one metropolitan region—the seven-county San Francisco Bay Area—from 1970 to 2012. We drew information from publicly available sources to capture the distribution and shifts in Bay Area postsecondary institutions. We began by generating a cross-sectional dataset to document the geographic locations and types of institutions in the region, as well as within-institution changes occurring during this period such as types of faculty employed, academic programs offered, and types of graduates. Because federal datasets (e.g., Integrated Postsecondary Education System [IPEDS]) are limited to institutions that participate in Title IV of the Higher Education Act (federal student financial aid programs) (Jaquette and Parra 2014), we utilized additional data sources, including the Bureau for Private Postsecondary and Vocational Education. To examine trends over time, we drew data from the Higher Education General Information Survey (HEGIS; from 1970 to 1985) and IPEDS (from 1986 to 2012) and generated a panel dataset. We also used sources such as the US Census and California Bureau of Consumer Affairs to identify a fuller range of postsecondary institutions in the Bay Area than the federal data sets included.

Across our data sources, we found that public systems were counted consistently and accurately but nonprofit and for-profit private postsecondary institutions were inconsistently represented in the data and frequently overlooked. In addition, programs that operate outside the confines of traditional postsecondary institutions, such as the dozens of "bootcamps" providing short-term training in computer programming around the San Francisco Bay Area, do not appear at all in the current postsecondary data (Lewin 2014). We address in depth the limitations of the extant longitudinal postsecondary data in appendix B.

We also employed quantitative data to examine both public and private postsecondary institutions in three geographic clusters: East Bay (Alameda and

Contra Costa Counties), Greater San Francisco (Marin, San Francisco, and San Mateo Counties), and South Bay (Santa Clara and Santa Cruz Counties). Within each cluster, we selected examples of four types of colleges: (1) a public four-year California State University (CSU) campus; (2) at least one public two-year community college; (3) a private nonprofit school (either a two- or four-year college); and (4) a for-profit college (see table 5.1 for a list of case study schools). Within these colleges, we focused primarily on four postsecondary programs salient to the Silicon Valley region: (1) engineering, (2) computer science, (3) biology/biotechnology, and (4) business administration. Longitudinal data were assembled on these programs and colleges.

Phase 3: Qualitative Research

The qualitative component of our study entailed document analyses, semi-structured interviews, and focus groups with a variety of stakeholders such as regional economists, college administrators, faculty, and program directors from a representative sample of postsecondary institutions and their industry partners in our three geographic clusters. We made use of information garnered from local media, and we consulted college administrative records, including selected Academic Program Reviews, which provided details regarding curricular and faculty changes in selected courses (see Gardner Center's website). We sought to understand in greater detail "who"—what types of colleges—attempted to meet the changing needs of regional employers—and "how"—what strategies and tactics colleges employed to adapt to industry demands. Throughout our qualitative data collection and analyses, we endeavored to address the effects of the larger historical context, including policy decisions and administrative structures, on the events and trends that we describe. All interviews and focus groups were at least 60 minutes in duration and were audio-recorded and analyzed for recurring themes and patterns (Miles and Huberman 1994).

Policy Directions and Implications

Our book concludes with a discussion of policy setting in higher education, including consideration of distinctive features that shape policy making in this arena. The lion's share of policy attention to education is attracted by the kindergarten through twelfth grade (K–12) system, with only episodic and targeted attention to the issues of higher education. Indeed, higher education appears to enjoy what has been described as a "politics of deference" by political actors, in the sense that substantial discretion is delegated by public agencies to academic administrators and faculty in shaping the course of its development (Doyle and Kirst 2015; Zumeta 2001). Also, until recently, the sector has largely escaped partisan infighting, although now such issues as support for for-profit schools and the terms for student loans have become battlegrounds for conservatives and liberals.

For many years, educational matters, including higher education, were left largely to the states. Only recently, and still rather haltingly, has the US Department of Education and related bodies begun to engage in more assertive guidance and control functions. Moreover, policy setting in higher education is fragmented, with some issues overseen by educational agencies and interests, others by labor, and still others by those concerned with veterans' needs and benefits.

California provides a complex and shifting context for the state's colleges. The policy environment for higher education is the product of historical compromises and incremental decisions. The current patchwork quilt is still grounded in the 1960 Master Plan that was the envy of the nation for many years. However, the division of higher education into three tiers—research universities (UC), state colleges (CSU), and community colleges (CCC)—has resulted in the unequal allocation of resources, as the top tier is favored while the lower and middle tiers receive inadequate resources and experience overly restrictive treatment. The tier approach also fosters policies that target only one layer of the system and neglect important issues of integration and coordination across levels, such as coherent regional planning.

California's population has grown rapidly during the past half century; it is also younger and more diverse than the rest of the country's. This growth has resulted in ongoing expansion in the numbers of students seeking a college education. State funding has been volatile and failed to keep pace with increasing demand, so that substantial numbers of qualified students have not attained admission during the past decade to either the UC or CSU systems, with the result that community colleges must attempt to accommodate them (Campaign for College Opportunity 2015). California's rate of degree attainment ranks twenty-third in the nation; only 39 percent of those over the age of 25 having obtained an associate degree or higher (Finney et al. 2014). The rate of degree completion is above the national average for the UC system, but well below for CSU and community colleges. Even though fees and tuition have risen recently, California remains one of the lowest cost public systems in the nation.

California relegates the control of nondegree, specialized training to another agency, the Bureau for Private Postsecondary and Vocational Education. Originally operating within the State Office of Education, it has been transferred to the Department of Consumer Affairs, but critics point out that the information it collects about the quality of these programs and the oversight it provides is inadequate. Although it is apparent that the functions of these programs overlap substantially with workforce training, there is little coordination with these state and national programs. In the San Francisco Bay Area, the demand from adults of all ages for postsecondary education exceeds the supply of low-cost educational opportunities in domains that are the key drivers of regional economic growth. State policy lacks the steering mechanisms to address these problems.

Competition and adaptation in the Bay Area's higher education field takes place within a boom and bust economy and rising and falling government funding. Since the 1970s, federal aid has plateaued while state revenues as a percent of total college operating expenditures have substantially decreased. One of the offshoots of the reduction in public funding has been the rise in for-profit colleges in this area. Regulators are struggling to find ways to encourage these new systems but to curtail their fraudulent practices and improve their graduation rates. More generally, the state has little knowledge of and exercises weak oversight over the entire range of private sector educational programs, from nonprofit private colleges to specialized vocational programs to for-profit entities.

Many other areas call out for improvement and reform. High on this list are the development of new metrics and methods for assessing both individual student and college performance, improved utilization and integration of online and in-class teaching and learning, standardization and verification of new kinds of credentials such as certificates and badges, increasing uniformity of course and credit offerings to ease transfer from institution to institution, and improved ways to assess workplace experiences and mesh them with educational credit programs.

More generally, the educational policy apparatus lacks information about or adequate mechanisms for intervening to correct some of the major fissures that transect and partition the field, such as:

- the gaps and shortfalls between K–12 and postsecondary colleges and programs,
- the differentiation and separation of the three major tiers of college education in the state,
- the isolation between the degree-granting colleges and the many other postsecondary programs offering vocational training, specialized training, and various types of certification and badges, and
- the separation of higher education programs from those devoted to workforce training.

Indeed, in many ways, the organization of the policy sectors mirrors the dysfunctional organization of the systems that are the target of its reform efforts.

As policy setters ply their trade, they would be well advised to be mindful of the many significant differences—in structure, mission, practices—that are represented in diverse organization populations making up higher education. Each has somewhat different interests and concerns, and no one policy initiative will be appropriate for all.

2

The Changing Ecology of Higher Education in the San Francisco Bay Area

W. RICHARD SCOTT, BRIAN HOLZMAN, ETHAN RIS,
AND MANUELITO BIAG

Higher education as a specialized sector has deep historical roots and provides an enveloping context for all colleges. Colleges have no choice but to participate in the wider field of higher education since various players in this field train its employees, set standards for their work, and provide guidance and governance of its activities. For many years, the field of higher education operated in a manner somewhat insulated from wider societal forces, although political interests and economic pressures have always exercised some influence. The latter have become stronger in recent decades, but entities within higher education continue to exercise substantial strength. We employ an organization field perspective to guide our discussion of this arena.

An Organization Field Perspective

The concept of organization field can be employed in multiple ways, of which we will emphasize two. The first was introduced by DiMaggio and Powell (1983: 48), who define it as "those organizations that, in the aggregate, constitute a recognized area of institutional life: key suppliers, resource and product consumers, regulatory agencies, and other organizations that produce similar goods and services." Like the concept of "industry," organization fields are often constructed around focal populations of provider organizations—in our case, colleges—but unlike the industry framing, organization fields are expanded to include other types of organizations involved in providing critical resources, services, or controls. The approach we employ combines the insights of institutional and ecological theory to highlight the importance of both symbolic and material resources in structuring social life. A second definition of organization field, emphasizing a regional focus, will guide our discussion in chapter three.

Martin (2011) reminds us that the concept of field had its origins in work conducted in the nineteenth century in electromagnetism and fluid mechanics and later in German gestalt theory in psychology, which emphasized that much of the behavior of an object or social entity is determined more by forces or influences operating in its environment than by its internal characteristics. Prominent approaches to organization fields place emphasis on relational or network systems—social structures involving the linkages and flows connecting organizations to similar or different organizations (DiMaggio and Powell 1983). But some approaches borrow from Bourdieu's (1971) work on cultural structures affecting social relations to stress the importance of symbolic processes that create common meaning systems and rule- and norm-based frameworks (Meyer and Rowan 1977; Meyer and Scott 1983). Together, these relational and symbolic frameworks account for much of the orderliness and coherence of social life (Scott 2014: chap. 8).

While external forces and controls should not be overlooked, more recent scholars have challenged the view that organizations are merely passive pawns in the face of environmental constraints. They point out that actors occupy different locations in social fields and that they bring varying resources and capacities to bear in seeking to protect their turf and interests. They also emphasize that while fields share some common meanings and relational structures, they also harbor competition and conflict, so that ideas conflict and interests diverge. In this altered version, actors are not simply blindly following scripts or conforming to pressures but are, to a variable extent, "agents" capable of independent, self-directed action (Fligstein and McAdam 2012; Lawrence, Suddaby, and Leca 2009). In our case, colleges are subjected to strong external pressures from regulatory bodies and shared normative frameworks, but they are also constituted to be independent agents who are expected to strategically pursue their interests.

Many fields are not settled but, rather, are contested terrains. Actors within fields compete for various types of capital: physical (e.g., property, monetary resources), social (e.g., friendship networks, alliances), and cultural (e.g., expertise, taste) (Bourdieu 1977). Within higher education, for example, competition until recently was based on "prestige" and cultural standing, but, as we will observe, over time it has come to be more about scarce resources and ways to increase drawing power. Within a field, pressures for conformity or change flow in multiple directions: down from broader cultural frames and social dominance frameworks, laterally from exchange and competition processes, and upward from resistance efforts, innovation, or the mobilization of suppressed groups.

As cultural and relational systems, organization fields host a variety of beliefs and normative systems. These are the elements that give orientation, guidance, and meaning to social life. Any complex organization field contains an assortment of these systems, termed *institutional logics*. Logics are sets of "material practices and symbolic constructions which constitute [a field's] organizing principles and which are available to organizations and individuals to elaborate" (Friedland and Alford 1991, 248; see also Thornton, Ocasio, and Lounsbury 2012). Among the conflicting institutional logics that we will explore are those between the liberal and the practical arts, the pursuit of prestige and academic standing versus profits, and views of education as a public versus a private good.

It must also be stressed that organization fields are always subsystems, being affected by actions in neighboring fields, as well as more macro societal sectors. Because of our interest in the connection between colleges and the Silicon Valley economy, we will be particularly attentive to effects of the regional economic structure on higher education organizations (see chapter three). But colleges are also greatly impacted by wider societal structures that contain differentiated and sometimes conflicting objectives that are reflected and refracted in their encounter with higher education, as we elaborate in chapter four.

Diverse Types of Colleges
College Populations

Organization ecologists have taught us to recognize the importance of addressing the questions: "Why are there so many (or so few) kinds of organizations?" (Hannan and Freeman 1989: 7). Like the biological ecologists who examine the origin and disappearance of species, organization ecologists study the diversity of organizations. All organizations providing higher educational service share common features and broad objectives but also vary in their *organizational archetype*—structural elements that embody different vocabularies of action (Greenwood and Hinings 1993)—and in their institutional logics. These *organizational populations* developed in different time periods and under differing circumstances, and they continue to bear the marks acquired at the time of origin: they are imprinted by the conditions present at the time of their founding. Thus, our approach is historical in several senses. Not only have we compiled a longitudinal data set and will consider throughout our discussion the value of being aware of changes occurring over time in the field, but we emphasize that "even when we observe a system at one point in time, we are seeing a cross-section of elements that are the residues of past processes" (Scott 1992: 169).

From the earliest period in our nation's history, centers for higher learning have flourished and multiplied. Beginning with a handful of religious-oriented colleges modeled on European counterparts, the number and variety of colleges have grown rapidly over more than three centuries. Over time, however, because of natural isomorphic processes and efforts by professionals overseeing higher education who established category systems, the field has converged around a limited number of forms. Because they are complex and multifaceted entities, organizations can be classified in multiple ways (see Ruef and Nag 2015). But to date the most influential effort to capture and assess the diversity of colleges was spearheaded by the Carnegie Foundation for the Advancement of Teaching whose Commission on Higher Education in 1970 created a set of categories that have been updated over time (Carnegie Classification of Institutions of Higher Education 2015). Although each has numerous subcategories, based on criteria such as locale size (e.g., rural, urban) and type of control (public/private, profit/nonprofit), the six major categories identified are:

1. Baccalaureate colleges (liberal arts baccalaureate degrees)
2. Comprehensive colleges (baccalaureate and advanced degrees)
3. Research universities (focused on more advanced degrees and knowledge creation; may also include liberal arts programs)
4. Community colleges (associate degrees and certificates)
5. Special-focus institutions (e.g., theology, medicine, practical skills)
6. For-profit entities (special-focus, associate, baccalaureate, and advanced degrees)

Three of these categories are what population ecologists would term "generalist" in orientation: community colleges, comprehensive colleges, and research universities. Generalists, by definition, operate in a wider range of environmental conditions (they occupy a broader "niche") because they pursue a variety of goals or missions and offer a diversity of programs (Hannan and Freeman 1989). To accommodate such variety, these organizations exhibit a highly differentiated internal structure—with numerous specialized programs and personnel focusing on a specific mission—and tend to be *loosely coupled*, developing mechanisms that reduce interdependence and the need for coordination or consistency across units (Weick 1976).

Relatively "specialist" populations include for-profit programs, liberal arts colleges, and special-focus institutions such as law schools. Their missions are more concentrated and, as a consequence, their structures—including staffing, curricular services, and support services—can be simpler and more highly coordinated. This implies that although generalists can outcompete specialists in diverse or rapidly changing environments, they can often be outperformed in any given field by specialized providers.

Given this taxonomy, we briefly discuss the major populations of organizations providing higher education.

Baccalaureate Colleges

These organizations are built around the model of the relatively small and self-contained liberal arts college. They were the first colleges to be established in this country, were closely modeled on their European counterparts, and they have continued to thrive. Some (e.g., Santa Clara University, Stanford University) have served as the nucleus for the creation of a research university, but others (e.g., Mills College) remain small and adhere closely to their original mission. They typically exhibit relatively high ratios of teachers to students and emphasize the importance of individualized instruction, such as tutorials and small seminars.

Another hallmark of these colleges is their emphasis on residential education. More than the other types of colleges, students in liberal arts colleges are likely to live on campus in housing provided by the school. These schools are also likely to offer medical, counseling, recreational, and social services. Students are expected to attend full time. They are, more than students in other types of colleges, immersed in a "total institution" (Goffman 1961)—an environment that envelops the student (client), defining who he or she is, and structuring the environment within which his or her life is lived. While the foregoing may be somewhat of a caricature, a considerable distance separates the educational experience of a full-time residential student engaged in a baccalaureate program from a part-time student pursuing a shorter program and living in the community.

Baccalaureate colleges have long been the bastion of liberal education, providing students a broad introduction to the humanities and sciences. These institutions place an emphasis on cultivating cultural sensibilities and civic virtues—producing educated, virtuous citizens. Education is viewed as a public good. However, we know that this is an overly idealized view in that many students attend college for other reasons, such as to participate in their athletic programs or be on the "party track" (Armstrong and Hamilton 2013). And, as discussed in later chapters, programs emphasizing career and vocational training are displacing the liberal arts even in colleges that have long championed them.

Comprehensive Colleges

These colleges began to emerge in the latter half of the nineteenth century as vocational and professional programs were attached to baccalaureate colleges. Colleges in the United States, compared to their European counterparts, were much more willing to incorporate these practical programs into their portfolio. The majority of these colleges are public, supported by states or large cities. They experienced rapid growth with the passage of the Morrill Land Grant Act—enacted in 1862 and expanded in 1890—whereby the federal government partnered with the states to "promote the liberal and practical education of the industrial classes in the several pursuits and professions of life" (www.law.cornell.edu/uscode/text/7/304). The growth of this form was also driven by the interests of professional associations, which were eager to connect their training programs to colleges and universities (Bledstein 1976). Examples of comprehensive colleges in the Bay Area include San José State University, San Francisco State University, and Holy Names University. San José State University began life as a normal school (teachers' college) and is the oldest public institution of higher education on the West Coast. As we will document, it has played a significant role in the development of Silicon Valley.

Research Universities

These colleges, modeled after the German universities developed in the last decades of the nineteenth century, adopt as their principal mission the role of knowledge creation and research training. This mission was rather quickly broadened and democratized within the American context to serve not just the esoteric arts of philosophy, theology, and science but to include the practical arts, such as engineering, agriculture, and business administration (Bledstein 1976). Some, such as Johns Hopkins University and the University of Chicago, were established *de nova*, but most grew out of connecting themselves to an existing baccalaureate college. Many experienced a major expansion in the mid-1950s as they partnered with the federal government to conduct basic and applied research and research training to support, first, military and then broader scientific and medical research (Berman 2012; Lowen 1997). The United States is unusual in its mix of research universities, in that both private (e.g., Santa Clara University, Stanford University) and public programs (e.g., UC Berkeley) are regarded as being in the top tier. Santa Clara and Stanford Universities both began as baccalaureate colleges, migrating from one organization population to a different one by adding professional schools and doctoral programs and creating research institutes.

Stanford University, a critical player in the origin of Silicon Valley, opened its doors in 1891. Founded by the former state governor and railroad magnate, Leland Stanford, it was modeled on Cornell, a combination private university and land grant college that placed strong emphasis on training in the practical as well as the liberal arts. Its first president, David Starr Jordan, was recommended to Stanford University by his mentor Andrew White, former president of Cornell. Both Stanford and Jordan were in agreement that Stanford, unlike colleges such as Harvard or Yale, would include practical subjects such as engineering in the curriculum from the beginning (Elliot 1937).

UC Berkeley, the largest campus within the University of California system, is also the largest public research university in the Bay Area. The university began instruction in 1869 in Oakland, but moved to its present site in Berkeley in 1873. It is well known for its strong math, science, and computer science programs, and for entrepreneurship activities centered in the Haas School of Business.

Community Colleges

This form appeared early in the twentieth century—the first, Joliet College in Illinois, was founded in 1901—and community colleges have grown rapidly throughout the past century. California was the first state to create a system of public community colleges, launching the program in the 1920s. Their biggest growth surge occurred in the 1960s so that by 1975 community colleges enrolled some 60 percent of all California undergraduates and by 2006, over 70 percent (Douglass 2010).

Unlike more traditional colleges, community colleges serve a broad array of missions, including remedial education, vocational preparation (they dwarf in scale any other institution in the provision of such training [Osterman 2010]), liberal arts instruction, transfer to four-year colleges, and general adult education. From the origins of this form, there have been debates about the relative priority of transfer liberal arts programs versus more applied vocational programs. Transfer programs were favored by faculty, virtually all of whom had been trained in comprehensive colleges or research universities, and for some time received priority, but in recent decades, applied, vocational programs have gained ascendance (Brint and Karabel 1989, 1991). The growing interest in applied programs can be attributed in part to changing mixes of students attending these colleges as well as to the changing logics of higher education.

In the aggregate, students in community college differ substantially from students served by traditional baccalaureate colleges, exhibiting higher ethnic diversity and lower educational preparation (Deil-Amen 2015). They are also more likely to be older, married, employed, and to attend part time. Many do not graduate but take a few courses to upgrade their skills or to improve their employment prospects. Faculty in community colleges experience less autonomy, are less likely to be tenured or tenure-track, and more likely to be unionized. Increasingly, faculty members are employed part time, work as adjunct faculty, and are either employed in industry or teaching at more than one college.

Special-Focus Institutions

Specialized educational organizations have been part of the wider field of postsecondary education from the very early days of the nineteenth century, but they have had a changing relation to the field of higher education. Often excluded, they have become more prominent over time because of policy changes, as described in chapter four. Almost always, their orientation is vocational and they range from short-term programs lasting a few weeks or months to free-standing professional schools offering advanced degrees extending over two to four years. Many are private and include a mix of both nonprofit and for-profit forms, but the latter are more prevalent. The types of training provided range widely from truck driving schools to computer programming—but the great majority of these programs are in beauty/cosmetology, health care, alternative healing therapies, language training, and theology. (Specific examples of specialized programs in the Bay Area are provided in case example 2.A.)

CASE EXAMPLE 2.A

SPECIAL-FOCUS COLLEGES AND TRAINING INSTITUTES

The Bay Area's higher education ecology is populated by a vast array of specialized institutions. These colleges are often overlooked in discussions of higher education although they serve a large share of students in the region. To illustrate the diversity of specialized colleges, we provide descriptions of four of these programs.

Palo Alto University (Palo Alto, South Bay)

Palo Alto University is a specialized institution accredited by WASC (Western Association of Schools and Colleges), offering training in the field of psychology. In 2012, it enrolled 876 students, more than 700 at the graduate level. It is also a nonprofit institution, governed by an independent board of trustees since its founding in 1975 as the Pacific Graduate School of Psychology.

Palo Alto University offers six degrees, two undergraduate and four graduate. Its two BS degree programs cater exclusively to transfer students who have completed the first two years of undergraduate study elsewhere. Most of these students come from three "partner institutions," including De Anza College, Foothill College, and the College of San Mateo, all community colleges. The institution's undergraduate division only dates to 2006. Palo Alto University also offers two master's degrees (an MA in counseling and an MS in psychology, both also available through an online program) and two doctoral-level degrees (a PsyD and a PhD in clinical psychology, both accredited by the American Psychological Association). According to its website, full-time undergraduate tuition and fees in 2015 totals $21,348 per year; a hybrid evening/online BS option totals $16,011. Enrollment, while still modest, has grown since the institution moved to a new campus in 2009; between then and 2012, the number of undergraduates nearly tripled and the number of graduate students grew by 40 percent.

Cogswell Polytechnical College (Sunnyvale, South Bay)

Cogswell Polytechnical College is one of the oldest specialized institutions in the Bay Area, dating to 1887, the year it was founded in San Francisco as a technical training school. Since then it has moved southward, first to Cupertino and then to its current campus in Sunnyvale. Recently, the school announced yet another move, this time to San Jose. Between 1979 and 2006, it also operated a campus in Everett, Washington.

Cogswell College offers a specialized curriculum described on its website as "the fusion of digital art, audio, game design and engineering." Eleven BA and BS programs are available, all focused on technical aspects of the "digital arts," including audio technology, animation, and video game design. The college also offers a one-year MA degree in Entrepreneurship and Innovation. Full-time undergraduate tuition and fees in 2015 total $16,160 per year. The college is WASC accredited.

In addition to its campus moves, Cogswell College has experienced volatility in its enrollments. In 2012, it enrolled 404 students (all but 12 of them undergraduates), but in the previous year it had only 288. According to the Integrated Postsecondary Education System (IPEDS) data, since 1975 it has never enrolled more than 512 at one time. Unlike most broad-access colleges, Cogswell is dominated by male students; women made up just 21 percent of the student body in 2012.

A-1 Truck Driving School (Hayward, East Bay)

Many specialized institutions in the Bay Area offer no degree programs at all, focusing only on vocational certificate programs. The private for-profit A-1 Truck Driving School, in operation in Hayward since 1997, offers courses preparing students for commercial driver's licenses. Its flagship program is a Class-A tractor-trailer operator training, which includes 40 hours of classroom training and 120 hours of field training (spread between observation and behind-the-wheel time). A-1's website reports that this is a four- to eight-week course, but also emphasizes flexible scheduling, including weekend classes.

IPEDS collects no information about A-1, and the school does not list any statistics about enrollment on its website. In addition, it does not provide any information about tuition, although an advertisement on the site offers a $1,000 discount. The website is enthusiastic about job prospects for the school's graduates, stating, "We are confident we can place our CDL program graduates in a driver position after completing one of our comprehensive programs," and noting that "truck drivers are one of America's highest paying professions." The site describes the school as "certified in the state of California" and contains language referring to certification by the Commercial Vehicle Training Association, a national trade association with restrictive membership. The CVTA website, however, does not list A-1 as a member.

Dev Bootcamp (San Francisco)

Among the newest categories of special-focus institutions are "bootcamps," highly focused technology-training programs, resembling immersion-based language programs. Dev Bootcamp, headquartered in San Francisco with additional campuses in New York City and Chicago, claims to be the oldest such institution in the Bay Area, although it only dates to 2012.

Dev Bootcamp describes itself on its website as providing an experience that "transforms beginners into full-stack web developers." There is only one program available. The curriculum begins with a nine-week online module introducing students to the basics of programming languages, including HTML, CSS, Ruby, and JavaScript. This is followed by a nine-week immersive program on campus, which is described as a 60–80 hour per week commitment. The curriculum couples short lectures with pair and group programming challenges. The last week of the program focuses on a final project: "building a useful and potentially marketable web application." Instructors simulate the conditions of working on a real-life programming project.

In its three years of existence, Dev Bootcamp reports that over 1,500 students have graduated from its program. The 2015 tuition for the experience, which also includes support for career preparation, is $13,950. The program is not eligible for any federal financial aid programs, including veterans' benefits. Dev Bootcamp has a selective admissions process with short essay questions and an interview, but applicants are not required to submit any prior educational credentials and do not need prior programming experience. The institution does not report any statistics about its admissions numbers or retention rates. Students who leave before completing the program can receive prorated tuition refunds.

As discussed below, official statistics and many widely used definitions and data sets exclude most of these institutions from the field of higher education. As data allow, we include them in our conception of the field because of our focus on educational programs connecting to the Silicon Valley region. Another important provider of educational services are the in-house training programs conducted by private companies such as Cisco Systems and Hewlett-Packard. Such programs play an important role in technical education (Carnevale 1993), but companies regard much of the data about them as proprietary. Moreover, they operate under different rules and norms than do mainstream colleges, so we have not attempted to include them in this study.

For-Profit Enterprises

Clearly, colleges in this category overlap in function with those just described, but we treat them separately because they are defined less by their academic programs than by their distinctive rationale and structure. As noted in our discussion of specialized colleges, for-profit programs have operated for many years on the margins of the field, and the great majority of them continue to provide postsecondary training in relatively specialized niches. However, during the 1980s, a number of them began to provide generalized college training at the two- and four-year level as well as offer advanced degrees. They experienced rapid growth during the first decade of the twenty-first century, moving from less than 3 percent of students enrolled in degree-granting institutions in 2000 to nearly 10 percent in 2010 (Snyder and Dillow 2012).

For-profits differ from more traditional educational programs in that they are designed to primarily serve the interests of their shareholders. Their focus on profits causes them to concentrate on strategies for growth and cost reduction rather than seek academic prestige or pursue the broader mission of liberal arts programs and educating leaders and citizens. The larger and more successful of these enterprises operate a number of colleges or branches and have adopted a corporate model of organizing. They typically serve nontraditional student markets—catering to minorities and older students—and offer highly structured and focused programs with few electives. Curricular decisions are centralized so that authority is concentrated in senior managers rather than involving faculty in significant decisions. The great majority of teaching staff is part time, and much emphasis is placed on marketing programs and gauging customer satisfaction. Some of the stronger programs also stress student services, including academic counseling, coaching, and placement (Tierney and Hentschke 2007).

Since most of their students are working full or part time, courses in many for-profits have been adapted to fit their needs. Many courses are offered online. For "resident" students, classes meet several times per week for four to six weeks. Students begin classes the day that they are admitted. Most classes meet in the evening. A full-time student is defined as a student taking at least one course. Admitted students are sometimes granted academic credit for previously acquired skills or work experience.

The University of Phoenix, DeVry University, and the Education Management Corporation (EDMC) operate several branches of their colleges in the Bay Area. EDMC owns the Art Institutes as well as other college programs.

Since the University of Phoenix was a pioneer in the field of for-profit higher education, we provide a brief account of its origins and history (see case example 2.B).

CASE EXAMPLE 2.B
THE UNIVERSITY OF PHOENIX

For two decades, the University of Phoenix has been the market leader in for-profit higher education. While Phoenix is now a national institution, operating in all 50 states, it had its origins in the Bay Area and is still an important, though shrinking, component of the region's higher education ecology.

The institution's founder and longtime CEO (nearly until his death in 2014), John Sperling, attended both the City College of San Francisco (CCSF) and the University of California, Berkeley, and eventually became a tenured professor of economic history at San José State University. While there, he developed a curricular model aimed at working adults, which he proposed to offer on a contract basis to colleges and universities through a private for-profit corporation called the Institute for Professional Development (IPD). After failing to interest San José State administrators in the program, he piloted it in 1974 at the University of San Francisco (USF). The Institute's first student cohorts were police officers seeking bachelor's degrees and public school teachers seeking master's degrees (Sperling 2000).

Through the 1970s, IPD expanded and signed contracts with more colleges and universities but quickly ran into complications, including a lawsuit from a partner college in Colorado and an FBI investigation into alleged bribery in the USF contract. In response to these regulative challenges, Sperling moved the organization to Arizona, where it would come under the jurisdiction of the North Central Association of Colleges and Schools. As part of the move, IPD changed its name to University of Phoenix and applied for accreditation as a stand-alone institution (Sperling 2000).

The North Central Association granted Phoenix accreditation in 1978, and the new university enrolled a small cohort of students in Arizona. Sperling's primary focus, however, remained in California, which offered reciprocity to institutions accredited in other regions. In 1980, Phoenix opened a campus in San Jose and followed shortly after with one in Southern California (Orange County). To protect their ability to operate these institutions, Phoenix formed a lobbying group called Accredited Out-of-State Colleges and Universities in

California to push for legislation ensuring continued reciprocity. Many years of effort finally came to fruition in 2003, with a bill that allowed Phoenix and similar institutions to advertise themselves as fully accredited within California (Walters 2013).

Throughout the 1980s and 1990s, the rapidly expanding university focused its educational mission on working adults. In those years, the minimum age of enrollment was 23 and recruitment targeted students transferring from other institutions with at least 60 units of course credit. During the 1990s, Phoenix concentrated on vocational programs, including nursing, computers, and business (Breneman 2006). The school focused on students whose employers offered to contribute to their continuing education; by the early 1990s, 85 percent of students had their tuition fully or partially reimbursed by employers.

Two developments in the 1990s fundamentally changed Phoenix's trajectory. The first, in 1994, was the initial public offering (IPO) of stock in the Apollo Group, the university's holding company. Critics have charged that the switch from private to public ownership led Sperling and other leaders to focus too much on expanding enrollments and not enough on the core educational mission of the school. According to John Murphy, an early administrator forced out by Sperling in 1997, following the IPO the school's graduation rate quickly fell from 65 percent to 33 percent (Murphy 2013). By the early 2000s, the university had dropped its minimum age requirement, began enrolling students directly out of high school, and developed a range of new associate's degree programs.

The second transformation—the introduction of online course work—had an even bigger impact. In 1989, Phoenix established a computerized education division headquartered in San Francisco to develop and administer distance-learning classes. With the spread of the internet in the late 1990s, this technology rapidly grew to offer asynchronous classes to students all over the world. While it maintained its brick-and-mortar campuses, Phoenix now had the means to circumvent any obstacles in jurisdictions that did not offer full reciprocity to institutions accredited elsewhere. However, a federal regulation established by the Department of Education in 1992 to crack down on correspondence-based diploma mills forbade the issuance of federal student aid to any institution that enrolled more than 50 percent of its students through distance programs. Since Phoenix was rapidly transitioning away from employer-funded tuition and toward self-funded students who relied on federal loans, this hampered their ability to grow the online program. However, in 2006, the so-called "50 percent rule" was repealed, setting the stage for explosive growth (Carnevale 2006). By 2010, the university enrolled 460,000 students, dwarfing every other higher education institution in the country.

The year 2010 proved to be the high-water mark for Phoenix in terms of enrollments but a low point for their performance record. A report in that year by the Education Trust found that the institution's nationwide six-year undergraduate graduation rate was only 9 percent, and for online students, 5 percent (Lynch, Engle, and Cruz 2010). According to IPEDS data, the San Jose campus had only marginally better numbers, at 13 percent. (By comparison, San José State had a 48 percent graduation rate in 2010). Throughout the past ten years, the university has also been implicated in multiple investigations over illegal student recruitment practices, resulting in nearly $100 million in fines and penalties (Blumenstyk 2011). Most recently, Phoenix has been eliminating programs that do not comply with the Department of Education's gainful employment rules, which went into effect in 2015. These factors, combined with waning student interest in continuing education due to an improving economy, drove total enrollment at the university down to 206,900 in May 2015, less than half its peak five years earlier (Blumenstyk 2015).

According to an interview with Stacy McAfee, the vice president of Phoenix's San Jose campus, the institution is currently retrenching and focusing more on its original mission of tight coupling with industry. Her campus is actively pursuing partnerships with employers like Oracle, Adobe, and the American Petroleum Institute and developing programs to teach industry-specific skills to current employees, who will receive tuition reimbursements. In addition, informants report that the institution is seeking to directly partner with community colleges, a turn away from its earlier practice of competing directly with them for students. Gregory Cappelli, the Apollo Group's current CEO, has confirmed that Phoenix plans to eliminate most of its associate's degree programs and will be instituting admissions standards for the first time in its history. The company is also showing some signs of diversifying from its longtime focus on degree programs; in 2015, it purchased a controlling interest in Iron Yard, a coding "bootcamp" that trains students in specific information technology skills in short courses (Blumenstyk 2015).

In their instructive review of for-profit schools, Tierney and Hentschke (2007: 50) point out three significant markers of their recent status. First, of the roughly 9,500 postsecondary institutions in the United States, virtually half are organized as for-profit schools. Many of these programs are vocational, awarding diplomas rather than degrees. Second, despite their great number, they make up less than 5 percent of all postsecondary enrollments. Third, they are currently the fastest growing segment of higher education. We add a fourth observation: over the last two decades, for-profit degree-granting colleges have experienced great volatility, with rapid rates of growth during one period, followed by high failure rates during another. We explore these matters in more depth in chapter five.

In sum, the United States possesses a mix of diverse types of colleges, all of which flourish today. Each type developed at a particular and different time in history, and each still reflects to some extent the conditions present at the time of its founding. These history-drenched features shape the distinctive competencies and liabilities of each college population.

The Uses of Diversity

Our summary sketch of the broad types of organizational populations that provide higher education in America reveals a multiplicity of forms with varying missions. The diversity of organizations providing higher education services poses severe challenges to those who work within the field, seek to utilize its services, and hope to change or reform it. Complexity is challenging, but it is also a major societal resource. The wide range of types of colleges represents a rich spectrum of approaches to providing educational services. As organization ecologists Hannan and Freeman (1989: 7, 8) point out,

> A stock of alternative forms has value for a society whenever the future is uncertain. A society that relies on a few organizational forms may thrive for a time; but once the environment changes, such a society faces serious problems until existing organizations are reshaped or new organizational forms are created [a difficult, uncertain, and time-consuming task]. . . . Organization diversity within any realm of activity, such as medical care, microelectronics production, or scientific research, constitutes a repository of alternative solutions to the problem of producing sets of collective outcomes. These solutions are embedded in organizational structures and strategies.

The existence of diverse organizational populations also provides a wider range of alternative career paths for employees and styles of education for students than does a more uniform system. And, importantly, it provides more entry points. Students may leave one type of program but subsequently be admitted to another. Failure in one system is not final, as is too often the case in more standardized and tightly linked systems. Such diversity in educational organizations has long been a distinctive hallmark of the United States and should be regarded as one of our greatest assets.

Colleges in the San Francisco Bay Area

We begin by reporting the size of college populations operating in the Bay Area. This region is made up of seven counties in northern California: Alameda, Contra Costa, Marin, San Francisco, San Mateo, Santa Clara, and Santa Cruz, and currently hosts roughly seven million residents encircling the San Francisco Bay (fig. 2.1). The region is anchored by four cities: Oakland in the east, San Francisco to the north, San Jose in the south, and Santa Cruz to the southwest. We first depict the numbers of colleges by population operating in 2012,

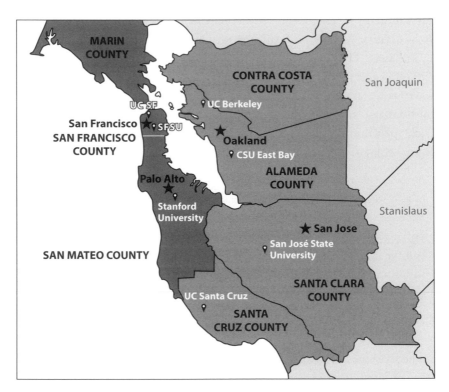

Figure 2.1. Map of San Francisco Bay Area with county boundaries and major cities and universities.

Table 2.1   Number of Colleges in the San Francisco Bay Area in 2012, by Highest Degree Conferred and Data Availability

	Multiple data sources	IPEDS
Public		
Research	4	4
Four-year and above	3	3
Two-year	23	23
Nonprofit		
Research	3	3
Four-year or above	78	37
Less than two-year	38	4
For-profit		
Four-year or above	36	13
Two-year	15	10
Less than 2-year	176	22

Note: Research-1 universities include Stanford University, Mills College, Santa Clara University, University of California Berkeley, University of California Santa Cruz, University of California San Francisco, and University of California Hastings. Data retrieved from IPEDS, California Postsecondary Education Commission, Bureau for Private Postsecondary Education, college websites, and other sources.

and then review trends over time in a subset of these colleges between 1970 and 2012 (where data are available).

In 2012, by our best estimate, the Bay Area was home to over 350 postsecondary institutions (table 2.1). The data reported in this table were garnered from multiple sources—including the Bureau for Private Postsecondary and Vocational Education, California Association of Private Postsecondary Schools, and individual college websites—to supplement the results reported in "official" educational statistics, specifically the Integrated Postsecondary Education System (IPEDS). Data from IPEDS are restricted to colleges that are degree-granting, accredited, and qualify for federal student loans. As a consequence, they include all degree programs operating under the auspices of public authority but omit many nonprofit and for-profit colleges and most postsecondary training programs. Even when data on for-profits are collected, they are often aggregated at the state or system level, making them less useful for regional studies. In particular, IPEDS lacks data on colleges operating programs under two years that result in diplomas rather than degrees. For a study such as ours interested in the interdependence of educational programs with the regional economy, this presents a major limitation. More generally, for all those affected by or concerned about the system of higher education, the absence of reliable information about the full range of educational providers poses a severe problem. (For a more extensive description of the various data systems, their strengths and limitations, see appendix B.)

Table 2.1 provides a comparison of the coverage of the primary populations of higher education providers by IPEDS with that obtained using a wider array of data sources. Clearly, IPEDS offers a restricted view of postsecondary programs in the field. While it captures all of the public providers, it excludes many of the nonprofit and for-profit programs: IPEDS omits about half of the nonprofit four-year colleges and almost 90 percent of the nonprofit schools operating less than two-year programs. Similarly, IPEDS omits about 70 percent of the four-year for-profit programs, 60 percent of the two-year programs, and about 90 percent of less than two-year programs. A review of these multiple sources indicated that the Bay Area contained 235 specialized institutions clearly focused on one field of instruction—more than 60 percent of the postsecondary population. The great majority of these were not captured by IPEDS data.

Research universities in the Bay Area include Santa Clara University, Stanford University, UC Berkeley, and UC Santa Cruz. The public four-year colleges are state comprehensive colleges: the three serving the Bay Area are California State University East Bay (CSUEB), San Francisco State University (SFSU), and San José State University (SJSU). This category also includes some private nonprofits, such as Golden Gate University. The public two-year colleges are all community colleges. Both nonprofit and for-profit programs that operate two-year colleges are specialized in mission. For-profit colleges that are four-year and above offer both academic and vocational training as well as selected professional degrees, such as business administration.

Figure 2.2 reports changes occurring between 1970 and 2012 in the numbers of degree-granting postsecondary institutions in the Bay Area. Figure 2.3 reports changes in enrollments for these institutions over the same period. Data represented in figures 2.2 and 2.3 are a reduced subset of those listed in table 2.1 because they only include those institutions that participate in the federal financial aid program. The population of four-year public colleges remained the same although enrollment slightly increased (fig. 2.3). These data support Brint and colleagues' observation that, unlike many other organization fields, in higher education, births and deaths in public colleges are comparatively rare (Brint, Riddle, and Hanneman 2006). Community colleges provide an exception: they increased in numbers until 1990 (fig. 2.2), and their

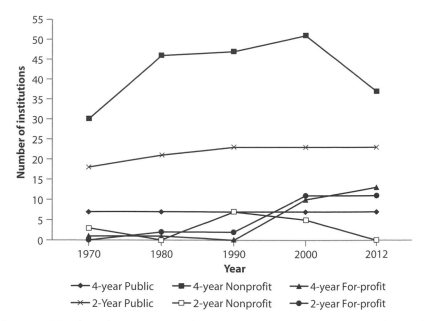

Figure 2.2. Number of institutions in the San Francisco Bay Area by level and control, 1970–2012.
Source: Higher Education General Information Survey (1970–85) and Integrated Postsecondary Education Data System (1986–2012).

enrollments increased sharply between 1970 and 1980 (fig. 2.3). Note that enrollments in community colleges are nearly twice those in all other types of colleges in the Bay Area, and would have been even larger had state funding kept pace with enrollment pressures (Callan 2014). The number of four-year public colleges, as noted, remained unchanged at three, and these have not expanded their facilities to keep pace with change. Space is often unavailable for qualified transfer students seeking to enter four-year programs, and many of the most popular programs in both two- and four-year colleges are "impacted"—unable to accept new enrollments (see chapters four and five). Figure 2.4 reports the extent to which the numbers of students academically eligible for the UC or CSU systems in the state exceed the combined enrollments of these systems. The differential has grown up to the present time.

Four-year nonprofit colleges increased in numbers from 1970 to 2000 but since then have declined. In a somewhat similar fashion, two-year nonprofit schools were few in numbers, increasing slightly in 1990, but thereafter declining. The most significant rate of growth is shown by the two- and four-year

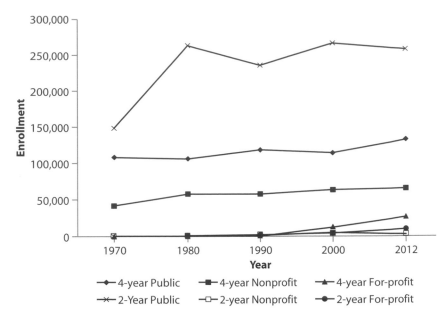

Figure 2.3. Enrollment in the San Francisco Bay Area by level and control, 1970–2012.
Source: Higher Education General Information Survey (1970–1985) and Integrated
Postsecondary Education Data System (1986–2012).
Note: We are missing data for the year 1990 for the four-year for-profit and the two-year
for-profit sectors; thus, the figure reflects 1989 data for these sectors. We are also missing
data for the 1980 and 2012 years for the two-year for-profit sector, so the figure represents
enrollment for 1981 and 2006.

for-profit schools, which have grown in numbers from 1990 up to the present
(fig. 2.2).

   While every region will vary somewhat in its specific make-up and dynamics, we believe our descriptive data are indicative of some of the more
general trends in the ecology of the field of higher education in the United
States. The field contains a half dozen or so different kinds of providers. Public
programs continue to be dominant, not in number of organizations but in enrollments. As highly institutionalized systems, their numbers have remained
relatively stable over a four-decade period. Two-year community programs
grew rapidly between 1970 and 2000, both in numbers of colleges and enrollments. They have become the primary vehicle in efforts to increase enrollment opportunities in this country. Four-year nonprofit programs continue to
operate the largest number of schools in the region, although their numbers
have declined since 2000. Two-year nonprofit schools never existed in large

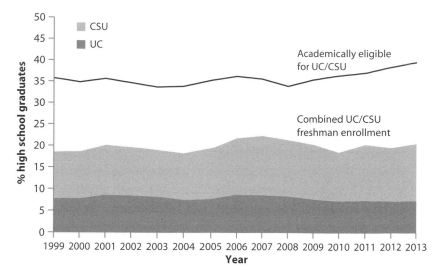

Figure 2.4. Numbers of academically eligible students for the University of California (UC) or California State University (CSU) system versus the combined freshman enrollment in these systems, 1999–2013.
Source: University of California, California State University, and California Department of Education.
Note: Used with permission from the Public Policy Institute of California and Jackson, Bohn, and Johnson 2016.

numbers and have declined since 1990. In contrast, both two- and four-year for-profit programs have increased in numbers since the 1990s.

Having introduced the major types of programs providing college-level educational programs, we turn now to consider the wider systems operating to support and regulate these entities, as well as the types of field forces at work (see also Scott and Biag 2016).

Organization Field Forces and Players

As noted, a sectoral organization field approach typically focuses on the providers of a particular type of goods and services, but also directs attention to those many other types of organizations that emerge to assist or control them. A staggering number and variety of these organizations exists in the United States, a clear sign that higher education is a highly developed, "mature" industry or field. A core proposition guiding organization field analysis is that *isomorphic* pressures will develop that encourage organizations of the same type to adopt similar structures and processes (Meyer and Rowan 1977). Within an organization field, these pressures include legal and regulatory frameworks that guide and constrain actors and actions, normative pressures conveyed through interpersonal and interorganizational networks, and cultural-cognitive frames that provide scripts for actors and templates for organizing (DiMaggio and Powell 1983; Scott 2014). The ability to closely resemble other organizations providing similar services is fundamental to winning acceptance and approval and obtaining legitimacy from one's peers and clients.

Isomorphic Pressures

Isomorphic processes occur at the field as well as at the organization population level. For example, as a vibrant and rapidly growing state, California was among the first states to develop a "master plan" for the organization and management of its higher education sector (see chapter four). While other states did not blindly copy the specifics of California's 1960 plan, they followed its lead in creating statewide structures to guide and oversee this rapidly developing arena. However, other forces surfaced that worked to dampen efforts to expand the higher education sector in California. In the late 1960s and early 1970s, when California became the epicenter of student antiwar activism and social protest, the negative feedback helped to lead the way for an ongoing retrenchment in state funding support for higher education.

Regulatory Systems

Whereas many Western democracies organize their higher education systems under a single, national system, from its origins the United States has opted for a mixture of public and private control arrangements and has delegated primary administrative control over higher education to the states. The federal government directly funds and oversees a few military academies (e.g., West Point), but exercises little to no direct control over most colleges. Agencies such as the Department of Education, while active in standard setting and funding some aspects of K–12 schools, provide primarily staff services to higher education programs such as student loan and accreditation programs. The federal courts weigh in on many issues confronting higher education, including diversity in student admissions and faculty and staff; equal protection issues such as financial support for women athletes; protection of free speech; separation of church and state; and freedom of information, including student and personnel records.

Public Systems. Most field-level administrative and regulatory controls occur at the state government level. Each of the 50 state systems contains a mixture of public universities, four-year colleges, and community colleges. States vary in mode of organization, but most operate under some type of state board of education with higher education programs often overseen by one or more specialized boards (Richardson, Reeves-Bracco, Callan, and Finney 1999; see chapter four for a discussion of the California system). Most states also establish some sort of regulatory system to oversee private postsecondary and vocational schools.

Private Systems. Many for-profit systems operate at a subfield level in the sense that they own and oversee collections of schools. While there are some public systems, such as the University of California, that operate at this level in the sense that they oversee multiple campuses and programs, private systems owned by for-profit corporations are much more likely to operate at a regional or even a national level than nonprofit or public systems. They grow, variously, by acquiring independent schools, developing new campuses, and partnering with public and nonprofit schools. In the Bay Area, the most significant for-profit educational corporations are the Apollo Group, which owns the University of Phoenix, DeVry, and EDMC, which operates the Art Institutes. The University of Phoenix currently operates three campuses in Hayward, Oakland, and San Jose and serves roughly 5,000 students. Until 2014, Corinthian operated nine campuses in the Bay Area, including Heald and Everest Colleges, serving more than 14,000 students, but after years of allegations that it altered grades, falsified job placement reports, and misled students in their marketing activities, the corporation has been forced to close its operations nationwide (Peele and Benedetti 2014; see chapter five).

For most for-profit systems, decision-making authority is centralized at the corporate level. Corporate board members, who have significant fiduciary interests in college performance, together with administrators make the strategic decisions about location and growth of campuses, as well as determine faculty composition and curricular matters (Tierney and Hentschke 2007).

A large number of associations have developed to connect specific types of private systems, both nonprofit (e.g., religious denominations and ethnic groups) and for-profits. While some of these exercise centralized administrative control, most primarily provide services to their members. Many of these associations—for example, the National Catholic Educational Association or the Southern Baptist Association of Christian Schools—include elementary and secondary schools as well as colleges and seminaries among their members. The most significant governance structures operating at the wider field level with respect to private schools are the accreditation agencies, described below.

Normative Systems

Normative controls "introduce a prescriptive, evaluative, and obligatory dimension into social life" (Scott 2014: 64). They typically operate through informal networks, communal ties, and occupational/professional connections. In modern societies, professional associations are particularly powerful sources of normative influence, as these groups set standards for their members and wider publics (Brunsson and Jacobsson 2000; Scott 2008). These associations are organized to operate at the national and, increasingly, international levels, and their authority in academic matters is considerable.

Meta-Organizations. A wide range of associations operates within the higher education field, some whose members are organizations and others whose members are individuals. They operate at regional, state, national, and international levels. Most of these provide mutual support for their members but at the same time exercise normative pressures on their members. Ahrne and Brunsson (2008) define "meta-organizations" as associations whose members are organizations. Meta-organizations are created to do many things, including helping members in their own operations, working to enable and enhance collaboration among members, attempting to regulate competition among members, developing standards and enforcing them, working to forestall or advance state and federal regulation, and engaging in lobbying efforts on behalf of the interests of their members.

Each of the major categories of colleges we have identified is associated with a national association to represent its interests. These are:

1. The Association of America Universities (AAU), representing about 60 major research universities
2. The American Association of State Colleges and Universities (AASCU), with more than 400 comprehensive state colleges and universities as members
3. The National Association of State Universities and Land Grant Colleges (NASULGC), representing about 200 public colleges and land grant state universities

4. The National Association of Independent Colleges and Universities (NAICU), with most of their 900 members being liberal arts colleges and comprehensive colleges
5. The American Association of Community Colleges (AACC), which contains over 1,100 community, technical, and junior colleges
6. The Association of Private Sector Colleges and Universities (APSCU), which represents about 1,500 for-profit institutions
7. The American Council on Education (ACE), which is a "peak association" attempting to coordinate and represent the interests of the entire higher education community of providers. On the one hand, it represents over 1,400 individual college members, and on the other hand, it attempts to coordinate the work of the five other major association (Cook 1998).

These "Big Six" associations (those listed above except for APSCU) operate offices in Washington with a full-time staff of administrators and lobbyists. Many hire nonprofit law, consulting, and lobbying firms to advance their interests. While they are joined by some common interests, the diversity of their membership and missions means that the issues that divide them are usually as many and as strong as those they share. As a consequence, higher education associations are not always effective in advancing their interests. For example, in the late 1960s to early 1970s period, these associations were unsuccessful in lobbying for federal support for colleges and universities. Instead, in 1972, with the passage of the Title IX Higher Education Amendments, Congress elected to provide aid to students rather than to institutions. Moreover, this aid was made available not only to traditional public college students but to students enrolled in vocational and technical programs, including those of for-profit institutions (Cook 1998: 26–27). This decision changed the boundaries of the higher education field in the United States to include formerly excluded proprietary programs (see chapter four).

Conversely, national educational associations—in particular, the NAICU—have successfully blocked attempts by educational reformers to collect and report "student unit record" information, which would allow the tracking of individual students throughout their college career to graduation and beyond (McCann and Laitinen 2014). This development has set back attempts to rationalize the system by taking into account the large amount of mobility of students between colleges.

Many colleges are umbrella-like systems that incorporate diverse and semi-autonomous components, such as professional schools and athletic programs. Associations have grown up to support and, in some cases, to oversee these entities. All of the professional schools, including medicine, law, business, engineering, and education are served by at least one, and more commonly a number of, associations. Examples include the Association of American Medical Colleges and the National Collegiate Athletic Association. In recent decades, these national bodies have been joined by international counterparts. Their influence on professional schools, their standards, practices, and curriculum, often exceeds that of the universities or colleges that house them.

Associations of organizations and their component units act as central nodes within networks of similar entities to convey information and know-how, standards and templates, socialization, and a sense of participation in a common enterprise. They both enable isomorphic pressures and disseminate innovative ideas and practices. In the words of Ahrne and Brunsson (2008: 43), "When meta-organizations are formed, organizations create a new order among themselves. If the included organizations have previously been in contact with each other, it has been another type of contact; they have been involved in an order other than that offered by a formal organization. [In forming of joining a meta-organization] . . . they have, in organization theory terms, constituted an environment for each other." In other words, meta-organizations or associations play a central role in the structuring of an organization field.

Disciplinary Associations. As Clark (1983: 29) reminds us, in addition to being a network of varying enterprises, "a national system of higher education is also a set of disciplines and professions." Indeed, disciplinary systems are increasingly transnational in structure and operation. The disciplinary associations are especially salient for schools in the upper tiers of the field—the research universities, comprehensive colleges, elite colleges, and a subset of the special-focus institutions. For faculty members in these settings, discipline trumps enterprise.

Abbott (2001, 2002) argues that the resilience of the academic disciplines within higher education rests in their "dual institutionalization": "On the one hand, the disciplines constitute the macrostructure of the labor market for faculty. Careers remain within discipline much more than within university. On the other hand, the system constitutes the microstructure of each individual university. All arts and sciences faculties contain more or less the same list of departments" (2001: 208–9). It is in this sense that the primary work of upper-tier colleges is defined and controlled by professional schemas and practices that penetrate into the core of the organization. In Clark's language, it is helpful "to recognize the great extent of crosshatching in academic systems. Such systems are first-class examples, written large, of 'matrix structures,' arrangements that provide two or more crosscutting bases of grouping . . . [in this case, by discipline and enterprise]" (Clark 1983: 31).

The disciplinary associations and their members assist colleges in overseeing the quality of their faculty appointments. When a position is to be filled or an incumbent faculty member is to be reviewed for promotion, colleagues within the academic unit seek letters of recommendation from members of the discipline external to the college. In addition, many colleges routinely enlist members of the wider discipline to serve on advisory and review boards to help them to oversee the overall quality and performance of their departmental programs. Because one's colleagues and peers exercise their controls based on shared expert knowledge and common norms, their control attempts are more likely to be treated as appropriate and legitimate.

Disciplinary controls remain strong in top-tier schools, but their scope is being reduced as colleges hire smaller numbers of tenure-track faculty and give higher priority to the solution of practical problems and workforce training. By their nature, these programs require interdisciplinary approaches. Unlike disciplines, which are governed by academics and only loosely coupled with one another, interdisciplinary teams require mobilizing efforts and coordination mechanisms, providing greater justification for the use of managerial controls that strengthen the hand of academic administrators versus faculty.

Faculty Unions and Staff Associations. A related category of association utilized by faculty members are labor unions. The emergence and growth of teacher unions was associated with the rapid growth of universities in the 1960s, and further spurred by reorganization and retrenchment policies that began to occur in the 1980s. Their growth was also encouraged by federal and state legislation that permitted public sector employees to unionize. By 1995, roughly 40 percent of full-time faculty in American higher education were represented by labor unions (Julius and Gumport 2003). Ninety-five percent of organized faculty are employed in public institutions, and about half of these are full-time employees (National Center for Collective Bargaining in Higher Education and the Professions 2006). The major organizing agents are the American Federation of Teaching (AFT), an affiliate of the AFL-CIO, the

National Education Association (NEA), and the American Association of University Professors (AAUP).

A range of other professional associations connects staff and auxiliary personnel to others of their kind. University presidents, trustees, admissions and financial aid officers, human resource officers, development personnel, accountants, alumni officers, athletic directors, coaches, counselors, controllers, diversity officers, community relations personnel, public relations officers, student residential administrators, librarians, information technology personnel—all of these and many other occupational groups gather to socialize, share information and experience, and advance their common interests.

As in our discussion of associations with organizational members, it is hard to overstate the impact on field structuring processes of associations whose members are individuals. These associations are dedicated to preserving and enhancing the personal and/or professional identities of their members, reminding them of their occupational investments and commitments, connecting them to individuals with similar concerns, and informing them of recent developments, both threats and opportunities. Such networks channel and reinforce traditions and standards, promote "best practices," and serve as sources of continuity and stability as well as innovation and change.

Accreditation Systems. Whereas in most countries educational accreditation is carried out by a governmental agency, in the United States until quite recently, the function of quality assurance has been performed entirely by private membership associations. The US accreditation process developed in the late nineteenth and early twentieth centuries as educators came to recognize the great discrepancies developing among schools and the need for standardization of curricula and requirements. (For details on the organization of accreditation bodies in the United States and a discussion of accreditation problems encountered by the City College of San Francisco, see case example 2.C).

CASE EXAMPLE 2.C

ACCREDITATION PROGRAMS AND CONCERNS

During the latter part of the nineteenth century, there was growing dissatisfaction and confusion among secondary educators around college admission standards and the preparatory courses students needed to complete before moving on to college (Shaw 1993) as well as what programs and standards colleges should adopt (Alexander 2012; Veysey 1965). In response to these issues, higher education leaders decided to regulate themselves rather than seek oversight and direction from the government. They worked together to develop a common set of admission standards and their administration (Alstete 2004). The two main branches of accreditation, which we still see today, were established during this period: institutional (regional) and programmatic (specialized) accreditation (Young, Chambers, and Kells 1983).

Regional accreditation associations began to develop in 1905, and as of today, six regional agencies operate to oversee the quality of colleges and other educational programs in the United States. All of the colleges offering baccalaureate and advanced degrees in the San Francisco Bay Area are accredited by the Western Association of Schools and Colleges (WASC). The association operates three commissions: one for elementary and secondary schools (K–12), one for community and junior colleges, and one for senior colleges and universities. Each is charged with setting minimal institutional standards, building institutional capacity, establishing quality assurance for third parties, and providing consumer information (WASC 2015). For example, in offering loans to students and veterans, the federal government relies on accreditation bodies to ensure that taxpayer support is going to students attending legitimate institutions (Glidden 1997; Goodwin and Riggs 1997).

In the early years of accreditation, programmatic or specialized accreditation was most active in the field of medicine, with the Carnegie Corporation and the American Medical Association's Council on Medical Education, taking lead roles in advocating for common standards in American medical schools. Today, programmatic accreditors review programs in a range of disciplines, including law, engineering, and business, among others. Accrediting programs is an important measure in facilitating smoother transfer of courses and programs among colleges and universities (Eaton 2009). Later, the Carnegie Foundation for the Advancement of Teaching was a key player in standardizing college-level education in the early twentieth century. When its trustees established a set of rules and regulations, which standardized the meaning and definition for colleges and universities, many colleges complied with these rules to continue receiving money from the foundation (Alstete 2004).

Concerns about Accreditation

Recently, the effectiveness of accreditation as a form of quality measure has been called into question as a growing number of students, families, and taxpayers raise concerns over the value, cost, and quality of US higher education.

Arum and Roksa (2011), for example, argue that accreditation has not always produced educational quality or improved academic rigor. Other critics contend that accreditation can be a burdensome, time-consuming, costly, and overly bureaucratic process with few returns to the institution or program (CHEA 2006). Additionally, critics of accreditation believe that the process inhibits innovation and competition. They argue that the structure of accreditation is resistant to change, hampers the ability of colleges to update programs to comply with industry demands, and restricts new providers of higher education from entering the marketplace (Burke and Butler 2012). Further, open-source microcredentials, including digital badges, certifications, and other skill identifiers are creating low-cost alternatives to recognizing and validating a student's subject matter mastery and competency. At present, the educational offerings of these newer providers are not widely regarded as credible among many academics, since acceptance of credits earned through these means by traditional accredited postsecondary institutions has not been successful (Smith 2013).

Accreditation and the City College of San Francisco

Accreditation is, at best, an imperfect governance system. Some of the issues raised by its operation are illustrated by the travails of the City College of San Francisco, a two-year community college, founded in 1935, that today consists of 11 campuses. It is the largest community college in the state, enrolling an ethnically diverse population of about 90,000 students.

In 2006, a review by the Accrediting Commission for Community and Junior Colleges (ACCJC), a part of WASC, required that CCSF "develop a financial strategy that will: match ongoing expenditures with ongoing revenue; maintain the minimum prudent reserve level; reduce the percentage of its annual budget that is utilized for salaries and benefits; and address funding for retiree health benefits costs" (Caroll 2006). The college attempted to address these concerns in a progress report, but in 2010 the commission warned that the college needed to deal with the problem of unfunded liabilities, in particular their retirement accounts. In 2012, CCSF was threatened by a loss of accreditation because of "tangled governance structures, poor fiscal controls, and insufficient self-evaluation and reporting" (Asimov 2015; see also chapter five).

Because CCSF was the largest college in California to potentially lose its accreditation, the problem was addressed not only by college administrators but by broader political mobilization. Protests came from citizens demanding that the city commit funds to reverse cuts to classes, programs, and staff (Bale 2013), and faculty unions as well as San Francisco's attorney filed lawsuits to force the ACCJC to maintain accreditation for the college, claiming that the accreditation body was "out of compliance" with proper procedures in its decision. Though CCSF remained in operation during this period, it experienced enrollment losses of about 15 percent (Emslie 2013).

In 2014, both the US Department of Education (DOE) and a Superior Court judge independently ruled that ACCJC had acted unlawfully and had exceeded its authority in deciding to end CCSF's accreditation (Bear and Brooks 2013; Fain 2015b). ACCJC has granted CCSF a two-year extension, to January 2017) to correct the remaining problems, and the board of trustees of the college has regained its authority over all aspects of the college, including its budget. ACCJC, in contrast, continues to draw criticism. A task force convened by the chancellor of the state's community college system, Brice W. Harris, concluded that ACCJC had exceeded its authority and often failed to respond to suggestions and criticisms (Harris 2015). And, in December 2015, a report by the DOE found that the ACCJC was out of compliance with federal regulations in 15 areas, and gave the commission six months to comply (Kelderman 2015).

At its best, accreditation stimulates a self-assessment, continuing-improvement process in which each college can undertake to review its own performance against a set of external professional standards. However, it is important to emphasize that accreditation in the United States is based almost entirely on measuring the quantity or quality of "inputs," such as adequate governance and institutional oversight of academic programs, evidence of learning resources (e.g., library holdings), adequate financial resources and accountability, adherence to generally accepted criteria for academic credit, expenditures, degrees offered together with course and credit requirements for each, and appropriate qualifications for faculty. Data on completion rates have recently been added, but there is no direct assessment of student learning (American Council on Education 2012).

Cultural-Cognitive Systems
Cultural and cognitive systems are the "shared conceptions that constitute the nature of social reality and create the frames through which meaning is made" by participants within an organization field (Scott 2014: 67). Educational

systems are strongly institutionalized in the sense that there is high consensus among practitioners and the public about the basic nature of schools and colleges—the centrality of "teachers" and "students," the shared conception of concepts like "curriculum," "course," and "credit hour," and the value of being a "college graduate" (Meyer 1977). Although knowledge and learning are amorphous concepts, there is remarkable uniformity within the United States and increasingly around the world in what constitutes the set of disciplines to be included in a college curriculum, and the broad requirements for graduation (Ramirez and Boli 1987).

Moreover, the field of higher education is the only one in which all of its core participants have been trained and socialized by an organization that is the same or similar in many ways to the one in which they are employed. While this assertion has rung true for most of the twentieth century, it is beginning to change. More and more college faculty members find themselves working in a different setting than that in which they were trained. For example, while virtually all faculty participants are trained in a research university or comprehensive college, a large number of them find employment in community colleges or more specialized systems. As a consequence, many attempt to reproduce their training environment, putting emphasis on the liberal arts transfer function, and resisting attempts to develop programs aimed at providing terminal vocational training (Brint and Karabel 1989).

To an unusual extent among all the various types of organization fields extant in modern society, the field of higher education, particularly its top tiers—research universities, comprehensive and baccalaureate colleges, and professional schools—has been controlled by normative and cultural-cognitive forces rather than regulatory sanctions or market mechanisms. These forces have dominated for many years and created a system of tiered colleges within which individual schools have been pressured to adhere closely to a limited set of legitimate models. The dominant institutional logic in the core programs has been to uphold the value of a liberal education: to expose students to the great traditions of literature and the arts, to prepare them for being effective citizens, to teach them how to express their thoughts clearly, to learn how to learn and how to think (Zakaria 2015). Until the 1970s, the field of higher education exhibited relative stability and harmony. More recently, however, competitive forces have come to the fore and the field has become more turbulent.

Competitive Pressures

Even though isomorphic pressures for conformity and uniformity remain strong, a counter set of institutional pressures is increasingly disrupting the reigning consensus on what colleges are and how they should operate. These forces stem in part from the widely shared belief in modern societies that organizations (including colleges) are constituted to be "rational actors": agents having a defined set of interests that govern their organizing processes and work activities (Meyer and Jepperson 2000). As organizations, colleges are expected to develop their own distinctive mission and to pursue strategic objectives that somewhat differentiate themselves from their counterparts.

Rather than following the assumption of early organizational field scholars, such as Meyer and Rowan (1977) and DiMaggio and Powell (1983), that organizations routinely conform to isomorphic pressures stemming from their institutional environments, later scholars such as Oliver (1991) argued that many would opt to take more strategic actions to pursue their own interests. Such actions range from avoidance and compromise to defiance and attempting to craft new rules and practices. Indeed, as noted in chapter one, all fields are made up of both established players—organizations who benefit from existing conditions—as well as actors who challenge the status quo, seeking ways to challenge current beliefs and practices. The clearest example of such a challenge in higher education is represented by the insurgent rise of for-profit organizations into domains formerly reserved for public and nonprofit forms. Less extreme instances of strategic action have also been described above as the various associations representing each of the college populations or the interests of some category of professional actors such as college presidents or financial analysts mobilize to represent, defend, and lobby for their interests at state, regional, and national levels (Cook 1998). But in addition to these field- and population-level strategic actions, each college also must balance the trade-offs between the comforts of isomorphic conformity and the rewards of seeking strategic competitive advantage.

From the beginning of their existence in the United States, colleges have been more broadly engaged in their societies than their European counterparts, more subject to political pressures and to the play of market forces (Washburn 2005, chap. 2). Confronted with higher levels of geographical mobility among the states, colleges have long competed with one another for financial resources, student enrollments, athletes, faculty, and coaching talent.

Beginning in the mid-twentieth century, colleges and universities began to embrace a conception of themselves as strategic actors, competing for federal funds that grew as a consequence of the military needs of World War II and then the continuing Cold War. To construct and implement their strategies, managers exercised more control as decision making became more centralized, faculty governance was replaced or augmented by administrative controls, and the ratio of administrators to faculty increased. Colleges and universities began to develop mission statements and engage in "branding" activities to highlight their distinctive features and programs (Hasse and Krücken 2013). More specifically, a growing share of colleges has elected to view their students as customers and devote more of their attention and resources to meeting the students' preferences rather than their "needs" as defined by faculty. As a consequence, college curricula have begun to emphasize vocational and preprofessional programs as against liberal arts or civic education (Labaree 1997). Institutional logics increasingly view education as a "private" rather than a "public" good.

Rating Systems

Because professionals work on complex and uncertain problems where good performance can often lead to bad outcomes, they have long insisted that they themselves, the experts rather than laypersons, should collectively determine the quality of their work. Educators have relied on peer evaluations and, more broadly, on reputation—the accumulation of evaluations of past performances—as general indicators of quality. They also readily adopt qualifications (of individuals) and capacity (of organizations) as useful indicators of quality. Attention is focused on *inputs* and *processes*—the training of faculty, the test scores of entering students—rather than *outputs* or *outcomes* (Donabedian 1966; Scott 1977).

Accreditation agencies, staffed by academics, have subscribed to the criteria preferred by the "producers" of educational services; mass media companies, in contrast, have developed ratings that favor the perspective of the *consumers* of education. *U.S. News and World Report* began to introduce rankings of colleges in the early 1980s, relying first on the assessments of college presidents, the experts. When these judgments were criticized as the biased opinions of insiders reflecting the views of "old-boy" networks, they introduced other measures, most of which in the beginning emphasized measures of inputs (e.g., average scores of incoming students, funds devoted to undergraduate education). Later, other measures were included to assess not educational capacity but amenities attractive to students.

Recently, however, ratings have begun to incorporate more output measures. *Business Week*, in its rankings of business schools, has relied on assessments by "customers"—both graduates and employers. Colleges and professional schools are increasingly being compared and evaluated in terms of how well their graduating students fare in the competition for jobs; now, starting incomes of graduates are reported. Surveys have been conducted of recent graduates and corporate recruiters to gauge their opinions about quality of training and preparation (Zemsky 2009). Other approaches have examined "student engagement," and still others have developed instruments to test student skills such as critical thinking and problem solving (Arum and Roksa 2015).

Research examining the effects of these rankings suggests that while they have affected the behavior of external audiences, such as student applicants and resource providers, their largest effects have been on the decisions and behavior of college administrators and faculty members. Many college staff have internalized the judgments reported by the external ratings and increased their efforts to improve the scores awarded by these agencies (Bastedo and Bowman 2009; Wedlin 2006).

Rising Costs

One of the most prominent features of higher education in recent years is the increasing cost to students and their families of obtaining a college degree. For more than two decades the cost of a college education in the United States has risen by an average of 1.6 percentage points more than inflation every year (*Economist* 2014b). Three important changes related to costs are directly associated with increased competitive pressures. Two of these occurred in 1972 when a set of amendments to the Higher Education Act (1) determined that federal aid should be distributed to students rather than directly to institutions; and (2) expanded the conception of the field from "higher education" to "postsecondary education." The first change compelled colleges to compete for students receiving aid, and the second made degree-granting proprietary schools, such as the University of Phoenix, eligible to admit such students. From this time, nonprofits were forced to compete for students not only with other nonprofits but also with for-profits (Carnegie Commission on Higher Education 1973; Peterson 2007; see also chapter four).

The third change has occurred as a result of declining fiscal support from the federal government and the states for public education. With the end of the Cold War and greater attention to rising deficits, the federal government has cut back significantly on research grants and contracts to higher education. These changes, however, have primarily impacted research universities. Of more importance for mainstream public colleges, funding from the states has dropped from averaging over 50 percent of public college revenues in the 1970s to under 30 percent in 2012 (Snyder and Dillow 2012). As noted, these declines were associated with the antiwar and other protests occurring on college campuses during the late 1960s, which alienated mainstream support for colleges. Colleges attempt to make up these differences by increasing tuition and fees and competing for students—including out-of-state and foreign students—who are better able to meet these higher costs. Increasing competition for students also indirectly contributes to rising costs as colleges attempt to attract students by expanding their offerings or providing more attractive facilities and amenities, engaging in an "arms race" to attract the best students.

In competing for students, colleges are more likely today to balance the academic qualifications of students with their ability to pay for their education, and, over time, have accorded heavier weight to ability to pay. Kraatz, Ventresca, and Deng (2010) document how these changes have been aided by a restructuring of colleges to combine admissions and financial aid offices into a single "enrollment management" department. This reorganization—promoted heavily by the American Association of Collegiate Registrars and Admissions Officers—has increased the salience of financial considerations in admissions decisions. We consider changes in public support for colleges in more detail in chapter four.

External Control and Supports: A Concluding Comment

This brief overview merely hints at the complexity of the contemporary field of higher education. The several populations of college organizations are each embedded in a similar but differentiated set of control and support relations involving a diverse array of organizations. While we have stressed the constraints imposed by regulative, normative, and cultural-cognitive systems, it is important to recognize the extent to which the higher education field is a contested terrain with different types of associations and agencies often working at cross-purposes to secure their distinctive interests.

Figure 2.5 depicts the maze of types of relations surrounding a college. This figure overstates the uniformity of the field because each college confronts a

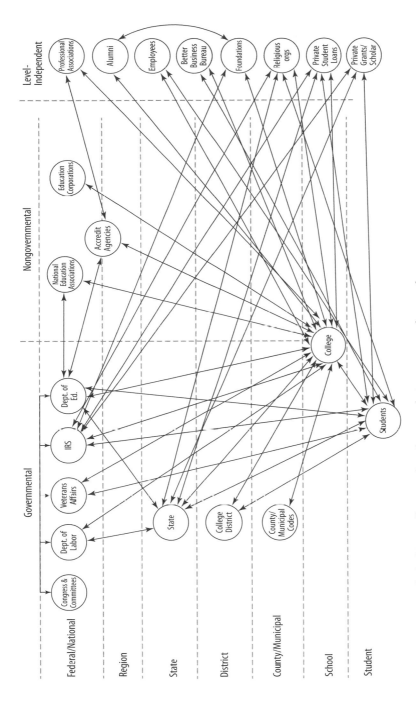

Figure 2.5. Networks linking colleges and students to their support and control systems.

specific configuration of constraints and supports stemming from the particular types of regulatory agencies or associations to which it is connected. For example, public colleges are strongly impacted by the requirements of the various federal bodies, but particularly by state policies. They are also highly responsive to the normative pressures stemming from the various professional and academic associations to which their faculty and staff are connected. In contrast, while for-profit colleges are also affected by the requirements of federal funding programs, their strongest links by far are with their corporate headquarters.

Still, the fact that the higher education system is still primarily a public entity helps to account for the remarkable stability that the field has displayed even in the face of substantial societal change and strong pressures for reform. The field has expanded and added additional players during the past half century but has not begun to experience serious restructuring. The changes encountered—in expanded numbers of students; increased multiplicity of student qualifications, needs, and interests; financial pressures; and challenges from online instruction—have been absorbed into and accommodated by the existing structure, at least up to the present time. New reform initiatives, new sources of funding, new regulations do not enter into a blank or empty field but must find their way into and through a myriad of existing rules, traditions, and vested interests.

This does not mean that no change has or will occur. The extraordinary growth of community colleges and the rise of for-profit colleges represent a very significant change in the ecological landscape of higher education. And, the development of online educational materials that are increasingly user friendly and sophisticated may still become the agent inducing some kind of major realignment in the provision of education. However, a more proximate source of pressure for change resides in the local economies within which all colleges are embedded. And if that economy is especially visible and vibrant, as is the Bay Area economy, then impetus for change is greatly increased, as discussed in chapter three.

3

The Regional Economy of the San Francisco Bay Area

W. RICHARD SCOTT, BERNARDO LARA, MANUELITO BIAG,
ETHAN RIS, AND JUDY C. LIANG

All organizations participate in multiple fields. In most cases, two are of particular significance: a sectoral field including organizations that provide similar services or products, and a regional field including organizations that develop interdependencies because they share a common geographical space. In the previous chapter, we described the effects on colleges operating in a specialized sector hosting multiple populations of providers and associated support and control systems. In this chapter, we begin to examine the ways in which colleges are shaped by the forces operating in the regions and communities where they are located and the collaborative relations in which they are embedded.

A regional focus has long been a staple of ecological studies as biological and social scientists have emphasized the importance of dependencies based on reliance on a shared locality and common pool of resources. Hawley (1950: chap. 12), one of the founders of human ecology, underlined the importance of regional or "community" interdependencies in social life. He argued that the geographic community calls attention to the *symbiotic* connections among unlike organisms or groups. Such differentiation gives rise to a division of labor in which specialists are able to collaborate to achieve common ends. More recently, regional economists point out that a concentration of similar firms in one region can create a pool of specialized labor and expertise that benefit workers, employers, supporting organizations, and the larger community (Moretti 2013). Regions host varying combinations of public and private actors and stimulate and support diverse modes of organizing.

While the focus of our specific research is on a particular region—the San Francisco Bay Area—our approach has implications for the examination of any regional economy. The specific mix of organizational actors—public, profit, and nonprofit—has important effects on hi-tech regional development, as studies by O'Mara (2005), Powell, Packalen, and Whittington (2012) and Saxenian (1996) demonstrate. Generally, all regional economies, including those focused on traditional manufacturing industries, wholesale trade, or financial services, present a distinctive organizational ecology that can be unpacked by the conceptual and methodological tools illustrated by our study. And all such economies in any modern society seek to harness the services of higher education to aid their efforts.

Organizations vary in the extent to which they are embedded in or oriented to the local region. The larger firms, such as Hewlett-Packard and Apple, are more attuned to national and even international connections and happenings than to their local environment. Similarly, college populations vary in terms of their connection to the local region. Research universities and comprehensive colleges are oriented to wider connections: state, national, and international. On the other hand, larger firms and colleges have a significant impact on the local regions in which they are located, so that the connections are somewhat asymmetrical. In contrast, community colleges and many nonprofits will be more integrated into their local context. For example, students for community colleges are drawn primarily from nearby communities. For-profits take pains to be highly responsive to local conditions and demands as they design their offerings. In short, region matters, but how much it matters varies by type of organization.

While the importance of locality has again gained acceptance, the rapid growth of information and communications technologies (ICT) allows intense collaboration among actors widely separated in space. New forms of postindustrial communities have emerged. Such communities work together to achieve both modest goals such as "making contributions in an ongoing way through the creation, exchange, revision, and recombination of shared knowledge" (O'Mahony and Lakhani 2011: 9) in domains ranging from software development to scientific problem solving, to developing common standards, such as in open-source communities. A large scholarly literature has developed to explore the advantages afforded to individual and organizational actors embedded in such communities (e.g., Marquis and Battilana 2009; Marquis, Lounsbury, and Greenwood 2011; O'Mahony and Ferraro 2007; Powell and Snellman 2004; Rosenkopf and Tushman 1998; Seely-Brown and Duguid 1991; van de Ven 2005).

Silicon Valley, in the heart of the San Francisco Bay Area, is the poster child of this type of community, which is both strongly dependent on local connections and prominently connected to wider, more distant collaborators (Saxenian

1996: 2008). Colleges are vital participants in this community, contributing crucial intellectual capital in the form of knowledge, and human capital in the form of educated and trained managers and workers (Berman 2012; Lowen 1997; O'Mara 2005). Of course, colleges also contribute directly to the economy of the regions in which they are located. They are often among the largest employers within their regions. Data collected by the Delta Cost Project reveals that the total annual operating and nonoperating expenses for the major colleges operating in the Bay Area are over $14 billion. These expenditures include the university's payroll, facilities, and construction and maintenance, equipment, supplies and services, funding for research, and spending by campus visitors (Siegfried, Sanderson, and McHenry 2007). Total expenditures for the University of California system in the Bay Area were over $7 billion in 2013; the California State University branches expended over $1 billion, and the community colleges in the area added more than $2 billion. Stanford University as a single entity recorded expenses of nearly $4 billion (Delta Cost Project 2013). Moreover, their presence attracts many other individuals and organizations who seek to benefit from the rich cultural and intellectual milieu they foster (Steinacker 2005).

Silicon Valley is also at the vanguard of a major shift occurring in modern economies as they pivot away from traditional manufacturing industries toward those that emphasize the creation of knowledge, ideas, and innovation. Moretti (2013: 49) estimates that jobs in the internet sector in the United States have grown by 634 percent during the past decades, more than 200 times faster than jobs in the rest of the economy. These jobs include not just those in science and engineering but also substantial parts of other industries, including entertainment, design, marketing, finance, and the medical sector. The Bay Area has even retained a strong hold on conventional manufacturing industries. A study by Helper, Krueger, and Wial (2012) reports that Silicon Valley had the country's highest manufacturing wages and the second highest concentration of production jobs in the nation's big cities. More than 160,000 people are working in Silicon Valley factories. Many have come back to the Valley not only because of rising wages in China but also because its proximity to hi-tech offers more control, quicker turnaround, higher quality, and more secure intellectual property. Equally important, growth within the innovation sector has a "multiplier effect" on jobs in other arenas, especially services. Those who work in restaurants, health care, hair stylists, real estate, and transportation all benefit from the spillover effect of a growing innovation industry (Moretti 2013). Hence, it is not surprising that the main metro areas of Silicon Valley, including San Francisco and San Jose, ranked in 2015 among the top big cities in the United States for job creation (http://www.newgeography.com/content /004165-the-evolving-urban-form-the-san-francisco-bay-area).

It is also the case that the Bay Area currently serves as a major motor for the entire California economy. The San Jose area saw a 6.7 percent increase in economic output in 2014 and the San Francisco-Oakland area a 5.2 percent expansion, compared to 2.9 percent for the state as a whole. Together these two Bay Area metropolitan areas generated 27 percent of the state's economic output in that year (Walters 2015).

Silicon Valley

Silicon Valley is a prime example of a region leveraging the advantages accrued from local clusters of knowledge and a complex mix of cooperative and competitive relationships (Kenney 2000; Lee et al. 2000; Saxenian 1996, 2000a). For the purposes of this study, the San Francisco Bay Area is made up of seven counties in northern California: Alameda, Contra Costa, Marin, San Francisco, San Mateo, Santa Clara, and Santa Cruz. This region, which encircles the San Francisco Bay and encompasses about 1,800 square miles (see fig. 2.1), contained approximately seven million residents in 2010. At the heart of this region is Silicon Valley, a global center for innovation that boasts a long line of impressive characteristics and achievements. For example, it contains the highest level of hi-tech manufacturing activity on a per capita basis of any major metropolitan area in the United States (Harris and Junglas 2013). The average level of educational attainment in the area is much higher than the state, with about 46 percent of the adult population having a postsecondary degree or higher, compared to 31 percent of California adults (Massaro and Najera 2014). The largest share of educated workers are immigrants, with over 60 percent of those working in engineering and other science-related fields born outside the United States (Handcock, DiGiorgio, and Reed 2013). Median income in Silicon Valley is much higher at $90,000 than the state at $60,000, and the country at $52,000. It needs to be emphasized that while the Bay Area economy is dominated by technology, other high-skill industries are also present, including health care and financial services. Although high-skilled workers are essential, other industries include, and depend on, many other types of workers having medium-level and paraprofessional training.

Even though the Bay Area is regionally circumscribed, its boundaries are highly porous and ever changing. As noted, it is a magnet for migrants who flow in and out of the region in varying numbers and from diverse origins. The area is also connected to collaborators who are far afield, often across the oceans (Saxenian 2008). Economic regions exhibit a complex ecology that affects all of their inhabitants, both individuals and organizations, but they remain open systems subject to the influences of a wider environment, as we describe in more detail in chapter four.

Silicon Valley: Multiple Reinventions

Although its international reputation dates from developments during the most recent half century, the roots of Silicon Valley were planted in the early decades of the twentieth century, and the region has undergone a series of transitions as varying industry clusters have arisen and declined (fig. 3.1).

Beginning in the early 1900s, researchers associated with Stanford University were pioneers in the *radio industry*, including radio-telegraphy and vacuum tube production (Sturgeon 2000). Early companies associated with these developments in the Valley included Magnavox and Litton Industries. A few decades later (first wave in fig. 3.1), the electronics industry was buoyed by *defense funding* and expanded into new areas as the United States prepared for and entered World War II. Companies fueled by this funding source include Lockheed Missiles and Space, Hewlett-Packard, and Varian Associates (Leslie 2000; Lowen 1997). With the invention of the *integrated circuit*, the principal new industry cluster emerging during the 1960s was semiconductors and other components for large computers. Major firms involved in these developments include Shockley Semiconductor, Fairchild, and its many offspring ("Fairchildren"), including Intel (Leslie 2000). Beginning in the 1980s, advances in technology led to the emergence of the *personal computer* industry, featuring the emergence of companies such as Apple, Silicon Graphics, and Sun Microsystems. Prior to the rise of this industry cluster, most of the professional jobs in Silicon Valley were technical in nature, but with the introduction of personal computers, consumer marketing and customer support became essential to the portfolio of many firms. Customers knew nothing about how to evaluate and compare the features of the products they were buying. Engineers remained important, but they were joined by marketing and business professionals (Kvamme 2000).

Following a period of slow growth in the early 1990s associated with defense cutbacks at the end of the Cold War, Silicon Valley reinvented itself again as it became a leader in the *internet* revolution, led by companies such as

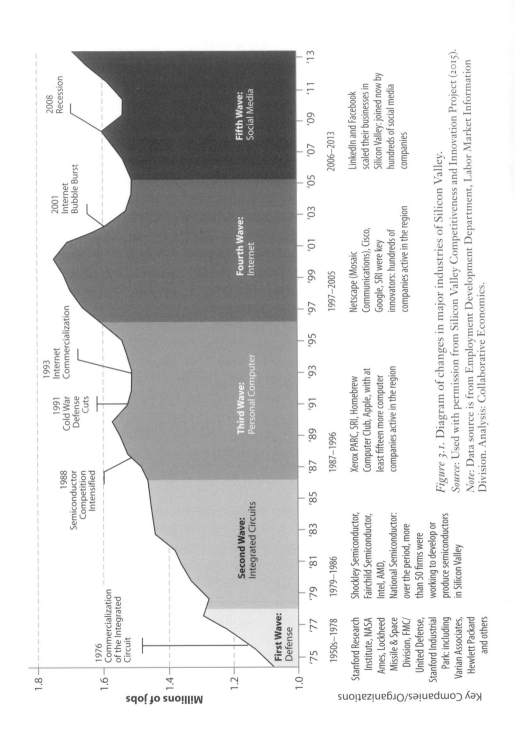

Figure 3.1. Diagram of changes in major industries of Silicon Valley.
Source: Used with permission from Silicon Valley Competitiveness and Innovation Project (2015).
Note: Data source is from Employment Development Department, Labor Market Information Division. Analysis: Collaborative Economics.

Netscape, Cisco Systems, and 3Com (Henton 2000). The development of this new industry was punctuated by the bursting of the "dot.com" bubble in 1997–2000, as too many inexperienced companies endeavored unsuccessfully to become e-businesses, attempting to take advantage of the opportunities associated with employing the internet for commercial purposes. Silicon Valley experienced an economic setback that continued into the early years of the new century.

However, during the first decade of the twenty-first century, a novel use of the internet was developed, giving rise to *social media*. A new set of providers employed the internet to harness interpersonal networks to share trusted information among friends and acquaintances but also to inform commercial providers on how to better serve, and commercially exploit, the members of these networks. Companies like Facebook and Twitter helped to lead this phase of the Valley recovery and fuel the continuing evolution of its economy. While this chapter's story is still unfolding, *bioinformatics* companies have emerged, connecting basic science in the universities with those of pharmaceutical companies and venture capitalists, to create value in genetics and medical arenas. Leading companies in these ventures include Chiron Health and Genentech (Powell, Koput, and Laurel Smith-Doerr 1996). Although the bioscience cluster did not grow rapidly at first in Silicon Valley, it has become the nation's leader with more than 500 companies (Henton 2000). And now it appears that *cybersecurity* concerns may fuel the next big wave of development. In short, the hi-tech economy of Silicon Valley has been able to reinvent itself more than a half dozen times during the past four decades, a remarkable exercise in dexterity and flexibility.

This history we have briefly summarized presents a nuanced picture of the multiple factors at work in generating and sustaining Silicon Valley. Other observers, however, suggest that the most telling factor in launching regional development was the partnership forged between the federal government and a university—in this case, Stanford (Lowen 1997; O'Mara 2005). O'Mara, in particular, argues that a number of university-centered economic development centers were the result of policies pursued by the federal government to shape the research priorities of universities, and to encourage their cooperation with private companies in the area. Many of these efforts targeted suburban universities and encouraged the use of land management policies that created attractive residential subdivisions, "taking into account where educated workers prefer to live" (O'Mara 2005: 2). The federal policies were carefully crafted to emphasize the role of the private entrepreneur, avoiding the appearance of an overreaching central state.

Stanford had largely missed out on federal funding during World War II in part because of the resistance to accepting federal funding by its conservative leaders, including its long-term member of the board of trustees, President Herbert Hoover (Nash 1998). Stanford's stance changed after the arrival of Frederick Terman, first dean of engineering and later provost of the university. As a student of Vannevar Bush of MIT, Terman recognized the value of partnering with the federal government. Working first with university president Donald Tresidder and later with Wallace Sterling, Terman cultivated federal research funds as well as local industrial partners to "build a community of technical scholars" in the area around Stanford (Gilmore 2001; Saxenian 1996: 22). Unlike numerous eastern universities, Stanford owned many acres of open land, on which it created a research park for "clean" industry, and unlike many eastern cities, Silicon Valley lacked a strong early industrial history or the constraints associated with growing up in the shadow of an old industrial economy. A pastoral setting provided the ideal location for a new industrial economy: a research university set within a suburban "green field."

Distinctive Features of the Silicon Valley Economic Region

Analysts have endeavored to ascertain the distinctive features of Silicon Valley—the "secret sauce" by which its continuous innovation and ongoing industrial renewal has been effected. The explanations have varied but we can discern some common themes.

The long-dominant, traditional industrial model is characterized by a field of large and stable companies that rely on vertical integration and specialized managerial control systems, offer long-term employment arrangements, and fund new ventures from surplus capital within the firm (Chandler 1977). In contrast, the Valley is marked by a complex mix of large and small companies, horizontal ties creating networks among firms, labor mobility patterns that move across companies (and countries) rather than up the hierarchy of a single organization, and independent venture capital funding. Unlike conventional commercial organizations that are run by executives and dominated by a "gray flannel suit and white-collar" culture, Silicon Valley firms are permeated by a strong engineering culture, emphasizing technical expertise, know-how, risk-taking, informality (blue jeans and t-shirts), a laid-back ethos, and disdain of hierarchy and rules. Scientists and engineers discovered that if they were willing to pursue their work from conception to prototype to production, they could reap financial rewards that had been reserved exclusively for top executives

(Lewis 2000). Although Silicon Valley is composed of a variety of independent organizations, many of them are amenable to regarding themselves as Lego pieces rather than as stand-alone firms. When needed expertise is not available within the company, it is as likely to be contracted out to another company as to be acquired. Because demand is constantly changing and product lives brief, supply chains are crafted into temporary networks that cobble together expertise from multiple companies (Saxenian 2000b).

An Unusual Mix of Organizational Actors

Manufacturing and Service Firms. Because in our regional inquiry we focus primarily on hi-tech industry, the "focal" population of interest are the hi-tech companies. Since the early 1980s, when the economic base began to turn from defense equipment and semiconductors to personal computers, an increasing number of small companies began to emerge alongside the larger, conventional firms such as Hewlett-Packard and Lockheed that had dominated the industrial economy of the region. A major force driving this change was the shift from a "proprietary approach" that locked customers into a single vendor of hardware and software services to an "open architecture" model encouraging new entrants to compete by developing differentiated products guided by a common industry standard. Sun Microsystems pioneered this approach and was quickly joined by many new start-up companies anxious to play—by simultaneously collaborating and competing (Saxenian 1996). Older and larger companies, such as Hewlett-Packard, who sought to join this emerging arena, were obliged to restructure, moving from a vertical to a horizontal organization, allowing discretion and flexibility for their operational units. Managers were compelled to learn new skills and develop novel business models to successfully compete. In addition to developing techniques for rapid decision making and turnaround systems, they needed to acquire the fine art of forging alliances, devising modes of outsourcing, collaborating with suppliers and customers, and mastering the norms of playing in a networked community (Nevens 2000). In addition, at least a subset of executives were enticed to move beyond mastering the skills of management to acquire the art of entrepreneurship: the ability to identify opportunities, calculate the risks associated with entering new businesses, and mobilize both internal parties and external partners around a new mission (Lee et al. 2000). In this realm, managers were joined and often led by engineers.

Central players in the organization field of the region's economy are the firms engaged in manufacturing and offering services in the hi-tech arena.

Primary players include establishments (e.g., firms) operating in two broad categories—information services and professional, scientific, and technical services. Data from the Bureau of Labor Statistics reveal that during the most recent decade and a half in the Bay Area (1998–2013), the number of establishments in information services has somewhat increased during this period with losses in the older areas of publishing and broadcasting being offset by greater increases in telecommunications and information and data processing services—from 760 to 1,650. The number of establishments in the area of professional services grew substantially, particularly in the areas of computer systems design (5,570 to 7,530) and management, scientific and technical services (3,252 to 5,980). In 2010, the percent of the Valley's gross domestic product (GDP) attributed to information services was 8.2 compared to 3.0 in the US economy; for computer and electronic product manufacturing, it was 9.2 compared to 6.9 for the United States; and 12.0 for professional, scientific, and technical services, compared to 4.7 for the United States (Randolph 2012).

Venture Capital and Law Firms. Other new or reinvented organizations include venture capitalists and Silicon Valley law firms. The rapid emergence of a high number of new start-up companies would not have been possible without the introduction of a novel mode of financing new ventures. Large business firms are able to rely on surplus capital to fund a new project group or new division; they are also better able to qualify for loans from conventional financial agents such as banks since they are perceived as a good risk. In contrast, small start-ups requiring substantial capital do not have access to these types of funds. A vital part of the innovation story that permeates Silicon Valley concerns the emergence of venture capitalists to meet this need.

Venture capital firms are professionally managed companies that loan money to new ventures in return for an equity share in the start-up. They are willing to assume a high degree of risk—most of the new ventures they fund will fail—because they believe that some of the firms they support will be highly successful, and they will participate in the increased value created. In addition to capital, they also provide other assets such as business and financial expertise. They frequently act as "coaches," consulting regularly with the start-up artists and engineers, who typically possess extraordinary technical know-how but lack business or managerial experience. Because the venture capital firms work with many companies, they can identify prospective collaborators or suppliers for fledging firms. Venture capitalists partner with their start-up companies on an ongoing basis, not only in the early stages of laying the foundation for the business but through the later, difficult phases of scaling up. Since the 1980s, the Valley has been home to the largest concentration of venture capital firms in the world (Florida and Kenney 2000; Hellmann 2000). The largest of these are concentrated in San Francisco and Santa Clara Counties (see chapter six).

An important related player is the Silicon Valley Bank. Founded in 1983 by a group of former Bank of America managers, this bank positioned itself as a broker connecting start-up companies with venture capitalists and professional service advisors. It has rapidly become a "major presence in the region," providing a range of services extending far beyond the purview of a conventional bank (Saxenian 2006: 31).

Unlike venture capital, law firms are not a novel professional organization. However, by learning how to provide a distinctive legal professional service, they have acquired unusual prominence in Silicon Valley. One lawyer, Larry Sonsini of Wilson, Mosher, and Sonsini, has been described in the popular press as "Silicon Valley's Secret Weapon," and as "the most powerful man in the Valley" (Suchman 2000). Sonsini and his colleagues were willing to work with engineers needing help with their start-up projects long before the larger law firms in San Francisco showed interest, although today most of these firms operate regional offices in Silicon Valley. Law firms in the Valley have learned how to serve not only as legal counselors but also as business advisers. Like venture capital firms, they participate directly in "deal-making," helping their young start-ups to identify appropriate partners for their enterprises as well as proffering general business guidance. As they move from firm to firm, they are able to discern general patterns or models for organizing that prove more or less successful, and they can impart this information to their clients. The firms also work with them to craft contracts between the start-up firms and the venture capitalists, serving as central intermediaries in this vital relation (Suchman 1995). Finally, they serve the conventional role of protecting their clients' property rights regarding new inventions.

Research Universities. Most accounts of Silicon Valley stress the central role played by research universities—Stanford University in particular, but also University of California, Berkeley, San Francisco, and Santa Cruz. As described above, under the guiding hand of Frederick Terman, dean of engineering (1946–55) and later provost (1955–65) at Stanford, faculty were encouraged to go beyond theoretical discoveries to develop working prototypes and then continue their testing and improvement by participating in founding a new company or partnering with an existing one. Faculty were urged to become entrepreneurs as well as scholars (Gibbons 2000). Also, early in the development of Silicon Valley, Stanford began to partner with companies such as Lockheed and Hewlett-Packard to provide advanced training to industry personnel. These types of relationships were advanced by the creation of the Stanford Industrial Park, established during the early 1950s to encourage new research and development (R&D) firms to locate facilities on Stanford land (Gilmore 2004; O'Mara 2005). Over time, other universities and colleges have entered into similar partnerships with companies, and UC campuses at Berkeley, San Francisco, and Santa Cruz currently play central entrepreneurial roles in the Silicon Valley economy (Kenney and Mowery 2014). UC Berkeley and Santa Cruz also operate extension programs that provide broad-access education to a wide range of adult learners (see case example 3.A).

CASE EXAMPLE 3.A
EXTENSION PROGRAMS OF RESEARCH UNIVERSITIES

Broad-access colleges are not the only higher education institutions offering explicitly workforce-oriented education in the Bay Area. The region's three research universities also offer programming to students interested in developing their job skills. They also often partner directly with industry to sponsor these students and create custom programs. The offerings described below are not designed simply as a public service; they are major sources of revenue.

The University of California (UC Berkeley) Extension

The extension program at UC Berkeley dates to 1892, with the launching of a series of free public lectures by university professors. In 1914, Berkeley's first "short course" was offered, in dentistry. Many others followed, offering certificates but not credits that could be used toward degrees. Instruction expanded beyond the Bay Area to four extension centers across California (Bakersfield, Sonoma, Sacramento, and Watsonville) (Stadtman 1967).

Statewide extension offerings continued to be administered out of Berkeley for several decades. After the full implementation of the California Master Plan, however, extension programs devolved to individual University of California campuses. In 1968, all direct funding from the state ended, and in 1982 most campuses ceased supporting extension programs out of their general budgets. This made extension work more sensitive to the demands of potential enrollees. Some scholars have criticized this new market orientation as shifting attention away from extension programs' original focus on the working classes and toward individuals from higher income backgrounds who already hold bachelor's or advanced degrees (Rockhill 1983).

Today, UC Berkeley Extension operates both online and out of three centers across the Bay Area: Berkeley, San Francisco, and Belmont (in San Mateo County). According to its website, its administration estimates its annual enrollment at 35,000 students spread across 1,500 classes. Although it is financially independent from the University of California, it still exists administratively under the UC Berkeley umbrella, with its own dean reporting to the executive vice-chancellor and having the same status as a professional school. In the 2013–14 school year, extension and summer programs brought in $735 million in revenue across the University of California (UC) system, while only requiring $279 million in expenditures. As of fall 2015, the most courses are offered at the San Francisco center (n=304), followed by Berkeley (n=244, including classes held on the UC Berkeley campus itself), and Belmont (n=55). An additional 170 classes are offered online. A plurality of these (24 percent) are in the field of business.

The University of California, Santa Cruz (UCSC) Extension

Extension programs at UCSC are almost as old as the institution itself, beginning in 1965. The first two decades of extension course work were dominated by noncredit "quality of life courses" offered to residents in the Santa Cruz area (which is much less densely populated than other regions of the Bay Area). That situation began to change in 1991, when the program began offering classes at a center in Santa Clara, in the heart of Silicon Valley. By 1999, they had additional facilities in Sunnyvale and Cupertino, as well as a joint facility on the Sun Microsystems campus in Milpitas. By 1999, 80 percent of UCSC's extension course work was in Silicon Valley (Benner 2002).

In recent years, the bulk of UCSC's extension students are already highly educated; as of 1999, 80 percent held at least one college degree, and 59 percent held two or more. Many were likely midcareer professionals, since 50 percent of all tuition was paid by reimbursement programs offered by students' employers. In the 1990s, UCSC Extension also worked to create training programs directly with a number of employers, including Sun, IBM, and Siemens Business Communications. These tended to be 12- to 18-month certification programs, often offered onsite at corporate offices (Benner 2002). The overall student population for UCSC Extension spiked during the dot-com boom, peaking at 52,000 enrollments in 2000–2001. Following the economic bust of 2001, these numbers dropped precipitously, necessitating the closure of most facilities and a 60 percent reduction in staffing (Desrosier 2010).

Today, according to budget records, UCSC Extension enrolls approximately 12,000 students a year. In fall 2015, 201 unique courses are available, some with multiple sections. Any domestic student can enroll with no prerequisites, although international students must hold a bachelor's degree. Courses are offered only at UCSC Extension's consolidated Santa Clara campus and online. The division's focus is explicitly on the information technology economy; its mission statement describes its core function as "advanced professional training that addresses the real-world needs of people who work and live in Silicon Valley."

More so than the UC Berkeley Extension, UCSC's program is heavily geared toward structured certificate sequences. Currently, 45 different certificates are offered. To receive a certificate, students must finish their course work with a 3.0 GPA; a 4.0 qualifies for a certificate with honors. Many of these courses are explicitly for midcareer professionals. For example, the Linux Programming and Administration certificate program offers two tracks: one for administrators who run Linux operating systems, and one for developers and programmers. While a few classes are offered online, students in this program must take some classes in person at the Santa Clara facility. In contrast, all of the course work required for the Technical Writing and Communication certificate (offered under the division of business and management) can be taken online. The ten-class sequence, which costs approximately $7,000, is targeted at a wide range of professionals, including engineering managers, experienced architects and designers, and "career changers."

Stanford University Cooperative Programs

The UC campuses are not the only research universities offering workforce-oriented course work to nonenrolled students. Stanford University's work in this domain dates to 1954, with the launch of the Honors Cooperative Program (HCP), described in chapter four. Classes in HCP, concentrated in the School of Engineering, catered primarily to working students pursuing graduate degrees. The program had tight partnerships with local employers, who often housed facilities for video instruction and tutoring at their workplaces. In addition, HCP provided the foundation for distance-based programs, including the Stanford Instructional Television Network and Stanford Online.

Today, extension programs exist under the umbrella of the Stanford Center for Professional Development (SCPD), but HCP continues to exist, offering eleven different master of science degrees to full-time professionals, eight of them in engineering disciplines. Tuition for these degrees is charged at the same rate as for ordinary graduate students and is almost always reimbursed by employers. Admission into the HCP programs is selective, and students must submit scores from the GRE exam and letters of recommendation.

SCPD also offers 28 "graduate certificates," ranging from Product Creation and Innovative Manufacturing to Financial Risk Analysis and Management. These are largely theory based, as opposed to SCPD's eight "professional certificates," which include IATA Aviation Management and Advanced Computer Security. Admissions criteria for certificate programs vary; some require a bachelor's degree and a brief statement of purpose. However, no testing or detailed applications are required. Graduate certificates typically require three to five classes, all offered online, and cost in the range of $15,000 to $20,000. Professional certificates are shorter, requiring less than 20 hours of online instruction or a three-day in-person course on the Stanford campus. Online courses cost between $250 and $1,300, while the in-person courses cost $3,000 and include access to online content. Finally, SCPD offers corporate education programs, which are custom-designed courses featuring Stanford faculty, offered online, at Stanford, or at a workplace. These courses tend to focus on broad themes, such as Entrepreneurship and Innovation, or Leadership. Stanford does not publish statistics on the number of students enrolled in any of these programs, or the revenue that they generate for the university.

If Stanford has served as the anchor tenant of South Silicon Valley, UC Berkeley has performed a similar role in Silicon Valley north. A recent study credits the faculty, students, and alumni of this university with spawning over 2,600 new firms since 1970, over half of them in the Bay Area (Bay Area Council Economic Institute 2014). With over 25,000 undergraduates and 10,000 graduate students, the university has become a leading center for entrepreneurial activities. Partnering with other Bay Area research institutions such as the Lawrence Berkeley National Laboratory and the Joint BioEnergy Institute, UC Berkeley has pioneered in areas such as designing the hard disk drive for early computers, helping the Bay Area to become the center of the relational database software industry, and developing the UNIX system. The university hosts or has spawned a variety of multidisciplinary institutions that support entrepreneurship, including the Lester Center for Entrepreneurship in the Haas School of Business, the Coleman Fung Institute for Engineering Leadership at UC Berkeley, and in partnership with UC San Francisco and UC Santa Cruz, QB3, the California Institute for Quantitative Biosciences, which supports basic research in the life sciences. UC Berkeley also participates with UC San Francisco and Stanford University in I-Corps: the Bay Area's NSF Innovation Corps, funded by the National Science Foundation to develop innovative ecosystems within universities to train the next generation of entrepreneurs (Bay Area Council Economic Institute 2014).

Powell and colleagues (Powell, Packalen, and Whittington 2012) describe the distinctive constellation of relations that linked Bay Area universities to local partners in the development of the biotech industry. Major university players included UC San Francisco, UC Berkeley, and Stanford, who early on spawned Genentech and Chiron to translate basic biological-medical science into clinical applications. These partnerships were substantially advanced by the support of venture capitalists in the area. The researchers note that "a striking feature of the Bay Area is the extent to which the commercial entities embraced academic norms [of publishing and collaboration] while the universities, particularly Stanford, came to venerate and support academic entrepreneurship" (Powell, Packalen, and Whittington 2012: 449).

Research universities also contribute to industrial and regional development through the creation of patents and licenses. Prior to the last decades of the twentieth century, it was assumed that knowledge produced in the university was publicly accessible to all, a part of the public good. However, the passage of the Bayh-Dole Act of 1980 allowed universities to own the knowledge produced by their faculty, including research funded by government grants. This legislation legitimated the commercialization of innovations and allowed universities to cash in on their intellectual capital. Stanford was at the forefront of this transition, being among the first universities to develop a technology-transfer office that provided a template for sharing revenues among individual contributing faculty members, their departments, and the supporting university, which has been adopted by other universities in the United States and around the world (Colyvas and Powell 2006).

Thoughtful observers have noted that the most serious threat posed by these developments is not the invasion of monetary incentives into university research or the attempt to combine basic with applied research, but the eroding of a central value in academic research: its open science culture. Academic norms have long celebrated intellectual freedom, collegiality, the open sharing of findings, methods, and data. In contrast, a proprietary culture restricts the sharing of knowledge and information, and hinders the unrestricted search for truth (see Dasgupta and David 1994; Washburn 2005).

State Universities. As noted in chapter one, while recognizing the vital contributions provided by research universities, this study attends to the role of broad-access universities and colleges in the Bay Area that, while not necessarily involved in initiating the Silicon Valley economy, have helped to sustain it over 40 years. For example, by the 1970s, San José State University trained as many engineers as either Stanford or UC Berkeley and became a leader in supplying not only engineering but also graduates in computer science and business to the Valley (Saxenian 1996; SJSU website; see case example 5.C). Later chapters will provide more detail on the role of state universities and other broad-access colleges in the Silicon Valley economy.

Corporate Colleges. A growing trend among companies, particularly hi-tech firms, is to create their own postsecondary training units rather than to rely on colleges to provide training for their employees (Meister 1998). A survey by the Boston Consulting Group found that the number of formal corporate universities in America doubled between 1997 and 2007, numbering then more than 2,000 programs. Companies operating such programs in the Bay Area include Apple, Cisco, Google, Hewlett-Packard, and LinkedIn. Among the advantages of creating such entities are that firms can tailor-make their curricula to fit company needs, instill corporate culture, and frame the program as a benefit provided to their employees. Moreover, it allows them to avoid the risk of losing a valued employee to a competitor when they use external executive MBA courses. A long-running survey of executive MBA courses by the *Economist* (2015b) reported that the number of employees who have their tuition paid for by their employers dropped from approximately 70 percent in 2005 to 40 percent in 2015. Clearly, these corporations prefer to maintain control over training their employees rather than source this function out to colleges and universities.

Corporate education is an important and growing component of the collection of entities that fuel the innovation economy, and one that in important ways displaces colleges seeking to offer training to current and future employees. However, because companies regard such data as proprietary, we were unable to obtain specific information on these programs for our study.

Other Research Entities. Other types of organizations vital to the success of the Silicon Valley economy include federal laboratories, such as Lawrence

Livermore, NASA's Ames Research Center, and the Joint Genome Institute, and independent and corporate labs, such as those operated by Hewlett-Packard, SRI, and PARC (Xerox). In addition, areawide organizations, such as the Bay Area Council Economic Institute, CIRIA (Center for Information Technology Research in the Interest of Society), and Joint Venture Silicon Valley, provide research-based expertise and sponsor forums for information sharing and coordination (Randolph 2012). The Bay Area presents a thick, dynamic web of idea generation and exploitation as well as information sharing among multiple types of private, professional, nonprofit, and public organizations.

Flexible Labor Markets

Hyde (2003) usefully points out that virtually all labor relations in the United States are based on employment contracts that do not require particular terms or benefits and can be terminated at will by either worker or employer. What is distinctive about labor markets in Silicon Valley is the extreme degree of volatility that permeates the employment relation. Flexibility resides at both the firm level—where mechanisms have been devised to rapidly increase or decrease the number and types of workers employed to fit changing market conditions (numerical flexibility)—and at the worker level—where workers are expected to shift from project to project, including cross-company projects, and to continually improve their know-how and skills (functional flexibility). Intense global competition and ever-shifting market conditions force workers to continually develop new skills and stay on top of technological advances.

As a result of increased competition among hi-tech firms, job changes are frequent in Silicon Valley relative to other regions (Saxenian 1996). While deeply committed to their work and colleagues, Valley employees operate more like "free agents" with no strong allegiance to any particular company. Because demand for hi-tech skills is strong and current demand outstrips supply, tech professionals, particularly those at higher levels, have a relatively easy time finding a new job in the region as they seek to maximize their pay and stock options (Harris and Junglas 2013).

The prevalence of contract workers in Silicon Valley is one of its signature features. Contract workers work alongside permanent workers, often doing the same type of work but without the job security or the benefits. While many, particularly those with lower skill levels, are forced to work as independent contractors, others choose to do so, citing the benefits of higher pay, heightened job autonomy, the opportunity to develop new skills and work experience, and the ability to take extended vacations. Some are employed by temporary hiring agencies as "W-2s," indicating that the agencies deal with employment taxes; others are "independent," some incorporating themselves as a business. Skill sets for contractors range from the very high end, such as those of developer-consultants and experienced experts, to the lower end, such as support technicians (Barley and Kunda 2004: chap. 3).

Compensation schemes in Silicon Valley also differ from conventional industries. Rather than press for high salaries and job tenure, Silicon Valley workers prefer to negotiate for a chance to receive stock options, being more than willing to trade job security and stable earnings for high risk accompanied by high rewards.

To an unusual degree, Silicon Valley makes use of temporary employment agencies and independent contractors. The Valley has twice the national percentage of its workforce employed by temporary agencies, and in Santa Clara County roughly 30 to 45 percent of the workforce is in a "nonstandard" employment relation. Nonstandard includes employment through a temporary agency, self-employed workers, part-time workers, and contract workers (Benner 2002: 38–45). While Silicon Valley is hungry for talent and continues to create jobs in the region, companies nevertheless look to lower labor costs through contractors and temporary workers, which, in turn, places workers in a more vulnerable position regarding job security and benefits.

Immigrant labor, in particular, has become a major source of skills for many Silicon Valley companies (Pastor, Ortiz, Ramos, and Auer 2012; Saxenian 2000a). In 2012, for example, approximately 36 percent of Valley residents were foreign born, compared to 27 percent within the state and 13 percent in the nation (Handcock, DiGiorgio, and Reed 2013). A large reserve of high-quality scientists, engineers, and tech professionals has been provided by China, Taiwan, India, and other countries, which has contributed enormous human capital to the region and provided linguistic and cultural know-how to encourage trade and investment flows between the United States and their home countries (Lee et al. 2000; Saxenian 2000c). These flows have been facilitated by the use of "specialty occupation" H-1B nonimmigrant visas, which allow workers to stay up to six years in the country (Salzman, Kuehn, and Lowell 2013). Inaugurated by the Hart-Celler Act of 1965, these special visas are granted to those whose skills are in short supply in the United States; the program has been extended by subsequent legislation in 1980 and 1990 (Saxenian 2006). STEM-related (science, technology, engineering, mathematics) occupations account for

two-thirds of programs intended to address skill shortages in the US workforce (Ruiz, Wilson, and Choudhury 2012). Top employers in the Valley re-
questing H-1B workers include Intel, Yahoo, Apple, and eBay.

As noted earlier, hi-tech communities are more likely to take advantage of
ICT improvements that support collaboration among distant players: individuals and organizations located outside the geographical boundaries of the
Valley. Saxenian (2006) calls attention to a new development: the role of Silicon
Valley in spawning a global network of immigrant-driven technology clusters,
as some foreign workers elect to return to their home countries to establish new
centers of innovation. The earliest of these were located in Israel and Taiwan,
but more recent centers have appeared in China and India. These developments
are fueled by a motley assortment of engineers and entrepreneurs—the "new
Argonauts"—who recognize the importance of "shared perspective, language,
and trust that permits open information exchange, collaboration, and learning
(often by failure) alongside intense competition" (35). These foreign workers
do not sever but retain their ties, connecting the new sites to collaborators—
both individuals and companies—in Silicon Valley. Evidence suggests that
both the Valley and the new centers flourish by a process better described as
"brain circulation" rather than "brain drain" (2).

Labor Market Intermediaries

Connections between higher education and employment (as well as between
workers and jobs) are increasingly mediated by a variety of *intermediary organizations.* Whereas in the past students relied primarily on the help of career
and placement centers in colleges, as well as on personal ties with family or
friends ("social capital"), labor markets in Silicon Valley are increasingly mediated by a range of new types of organizations. As Benner, Leete, and Pastor
(2007: chap. 3) point out, these organizations serve three major market functions: They facilitate "market-meeting," taking jobs as given and matching
workers with employers; "market-making," including affecting the quality and
distribution of jobs, providing worker training, acting as legal employer, and
engaging in advocacy; and "market-molding," activities such as pre-employment
and vocational training, facilitating information flows, and networking opportunities for workers and employers. Of course, these organized arrangements
supplement, rather than replace, a vibrant array of informal interpersonal networks operating in the Valley (Castilla et al. 2000). We describe below some of
the major types of intermediary groups.

Temporary Staffing Agencies

These temporary or "temp" agencies compile information on job openings and persons seeking jobs, negotiate matches, and place workers. In return, they charge clients a margin, based on their pay rate. Temp agencies often become the employer of record, withholding the workers' employment taxes, including federal, state, and local income taxes, social security taxes, and performing other administrative tasks (Kalleberg 2000). Employers benefit not only by escaping these administrative responsibilities but also forego the costs of recruitment, training, and terminating employees. Moreover, they are not required to provide temporary workers with fringe benefits such as life insurance and contributions to pension funds (Barley and Kunda 2004). Temp agencies serve workers by allowing them to test out a new job before making a longer-term commitment. However, such employees experience disadvantages, with lower job security and take-home earnings, reduced access to benefits, and being hardest hit during economic downturns. These problems fall disproportionately on women and minorities, since a larger share of them are employed through these agencies (Giloth 2010).

In Santa Clara County, in the heart of Silicon Valley, the number of temporary workers tripled between 1984 and 2000, growing over twice as fast as the overall labor force. Some agencies specialize in types of workers served (sales or technical, high- or low-end skills), while others are generalists. The largest temp agency in 1997 was ManpowerGroup, which operated more than 20 offices in the Valley (Benner 2002).

Web-Based Intermediaries

The high information costs associated with rapid turnover of personnel for both workers and employers have been eased by the growth of for-profit web-based agencies. Intermediaries, including large online job boards and social media sites, such as CareerBuilder, LinkedIn, Monster, and Craigslist, help to address this gap by aggregating, packaging, and selling information about job seekers and vacancies (Cappelli 2001; Gutmacher 2000; HR Focus 2000; Marchal, Mellet, and Rieucau 2007). Thousands of these sites, such as dice .com and LinkedIn, have developed in the past ten years and utilize a wide range of structures and business models. For example, Monster.com receives its revenues primarily from fees paid by companies for listing job openings. LinkedIn, the largest professional social network site, which started in 2003, now has over 300 million members in more than 200 countries (LinkedIn

2015; see case example 3.B for an account of these web-based employment agencies).

CASE EXAMPLE 3.B
LINKEDIN AND OTHER ONLINE JOB PLACEMENT SERVICES

Increasingly, communications are becoming more digital and socially networked. This is especially true when it comes to linking workers with jobs. As leading industries continually utilize the latest innovations in technology to attract qualified workers, colleges and universities are unable to keep up with these rapid changes—especially as their budgets have declined. Most colleges and universities are looking for ways to trim services, programs, and expenses, and we found that career centers and career-related services are among the first to be cut. Although these centers remain in many postsecondary institutions, the resources and services they are able to provide tend to be antiquated and less aligned with the needs of current students. As such, the burden of navigating the world of work often falls primarily on the student with little support from the college career center. Today, college students are increasingly using social media websites and taking the initiative to construct their own resumé and build their own personal network.

Online platforms such as LinkedIn and Monster.com are not only effective and efficient, but ever more necessary for both job seekers and employers to successfully obtain a job or hire an employee. Our interviews with postsecondary administrators, industry leaders, as well as college students, reflect current trends and suggest that LinkedIn, Glassdoor, and Monster.com are the most widely used professional social media websites.

As described by the *Economist* (2014c: 51), LinkedIn "is an online contact book, curriculum vitae and publishing platform for anyone wanting to make their way in the world of work." Founded in 2012, LinkedIn is the world's largest professional online network, with millions of people using it to network with others as well as to find and post potential job opportunities. In addition, LinkedIn has recently moved into the higher education space, "building an array of new tools for students in their search for a college and for a career" (Selingo 2015). In 2014, LinkedIn acquired Lynda.com to help fill critical skills gaps in the workforce. "A person could be job hunting on LinkedIn, realize they're missing a certain skill for a position, and learn that skill by watching a Lynda course" (Wong 2015). LinkedIn "wants to change not only the business

Table CE3.B   Top Ten Utilized Websites for Employment Searches (2015)

Company	Date founded	Approximate number of unique users per month	Main functionality
LinkedIn	2012	40 million	Professional network
Monster.com	1999	34.4 million	Job search and other career resources
Glassdoor	2007	26.5 million	Job information and comparisons
SimplyHired	2003	9.1 million	Job search and listings
Indeed	2004	3.7 million	Job search and listings
Dice.com	1990	2 million	Tech job search
CareerBuilder	1995	1.8 million	Job search and listings
Linkup	2005	1.7 million	Job search from company websites
Idealist	1996	1.4 million	Nonprofit job search
USAJobs	N/A	4.6 million	Federal job search

Source: jobsearch.about.com

of recruiting, but also the operation of labor markets and, with that, the efficiency of economies" (*Economist* 2014c).

Glassdoor is another widely used professional website, especially among students. Glassdoor was launched in 2008 and is headquartered in Sausalito, California. Unlike other online job boards or professional social networking sites, Glassdoor is "where people anonymously rate the places where they work or have been interviewed" (Doyle and Kirst 2015). It ensures that its employees' reviews are genuine by verifying and screening through e-mail addresses. In fact, approximately 15 to 20 percent of the content is turned away from Glassdoor due to failure to meet the company's guidelines. Glassdoor also provides other information helpful for job seekers, including company-specific salary reports, sample interview questions, and reviews of companies' benefits.

Founded in 1999, Monster.com has been one of the most visited employment websites in the nation. Its main utility is for employers to post job opportunities and for job seekers to find the job listings to match their skills and locations. Monster.com also focuses on partnerships and alliances to expand its reach for employers and job seekers. In late 2014, Monster.com partnered with the US Cyber Challenge to "build a community of cybersecurity professionals and verified talent pool that government and private sector employers can tap to fill positions in this critical field" (Boyd 2014). Knowing how to best use these and other online tools is critical for students to find employment and other opportunities (e.g., internships). A manager from Hewlett-Packard stated, "We force every student who wants a position to apply online . . . you can't show up at my table and give me a resumé." This sentiment, which is shared by many companies, calls into the question the ways in which colleges and universities currently prepare their students for today's labor market. Our interviews with students suggest that their college has played a minimal role in their job search. One recent college graduate, who now works at Cisco told us, "I didn't rely that much on my institution. I mostly relied on applying for jobs online and other resources and the different networks that I had . . . most of the opportunities that I got are from my network and not through me applying for jobs or anything." Another student who also is interning at Cisco described his job search process this way: "I would network a lot and if I have an interview with the company, I would try to talk to as many people at that company as I could." In addition to knowing how to use the professional online websites, students increasingly need to build a strong network. One student remarked, "Networking was absolutely the main method I used . . . the most effective approach I found was really talking to people and kind of getting a pretty good idea of what companies are looking for in candidates and also putting myself out there all over the web so recruiters can find me and really looking through the different websites."

Although colleges and universities still host job fairs and workshops on interviewing skills and resumé writing, our interviews indicate that few are providing supports necessary for the realities of today's job market such as how to navigate LinkedIn or other online sites, build professional networks, or reach out to industry professionals. Students commented on how career centers could improve on the resources and services they provide. One student said, "career centers don't give you any contacts . . . they can be more helpful by making themselves more available instead of just sending out emails." Another student commented that "career centers should teach students how to use social media to apply for jobs . . . maybe they can offer workshops to help people make their resumés stand out or the key words to put on your resumé."

When asked how students were getting connected to job opportunities, college administrators and faculty rarely mentioned their career center or counseling system; rather, they relied on their own professional networks to support students. As one administrator expressed, "Yes, I connect my bio medical students to jobs. I've got my claws into them. I have my contacts in industry, whenever there is an opening they let me know and I will send it out to my students. And now the alumni are working and they will send out a position that's open and have students send their resumes to them. And I will hand carry them. In biomedical engineering, I have been working to develop this kind of network but I don't think it's universal. So it needs to be developed at a higher level."

In sum, college students today are less likely to rely on their institutions to help them look for and secure employment. As LinkedIn's cofounder Reid Hoffman writes, "Both company and employee need to look outward toward the overall environment in which they operate, especially when it comes to networks. Companies have to understand the employee's broader place in the industry, while the employee should realize that his professional network is one of the key assets that can boost his long-term career prospects" (Hoffman, Casnocha, and Yeh 2014, 97–98).

Membership-Based Intermediaries

These organizations include professional associations, guilds, various types of unions, and immigrant associations. They help their members in a variety of ways, creating job postings, fostering networking opportunities and skills (including language training), and building learning communities (Benner 2002: chap. 5; Saxenian 2000a). Unions employ collective bargaining arrangements to improve employment conditions such as wages, benefits, and codes of conduct. To confront the volatility of economic conditions in the region and the demand for specialized skills, membership associations help to ensure that their members are up-to-date on current developments in their field and have the requisite skills to cope with a highly competitive job market. They also provide their members with professional training, information about wages, and assistance to acquire skills, including resumé building and negotiation strategies. These membership-based groups foster social networks that can help advance workers' careers (Saxenian, Motoyama, and Quan 2002).

Immigrant associations have been particularly active and influential in Silicon Valley. The first immigrant networks originated because of "shared educational and professional backgrounds, as well as common culture, language, and history" (Saxenian 2006: 49). Among the first to develop were informal associations among the Iranian, Israeli, and French professionals in the 1970s, to be followed in the 1980s by more formal groupings of Chinese and Indians, and still later by Hispanic, Korean, Taiwanese, and Vietnamese associations. Saxenian lists more than 30 such associations active in Silicon Valley (Saxenian 2006: appendix A).

Public-Sector Intermediaries

These organizations include a variety of federal and state workforce development programs that seek to find opportunities for disadvantaged workers and provide them with training and assistance in acquiring skills and know-how to enhance their employability. Unlike their counterparts in the private sector, public-sector intermediaries focus primarily on serving a population of minority and disadvantaged workers and are strongly shaped by state and local policy dynamics. In their most recent manifestation, these programs operate under and are largely funded by the Workforce Investment Act of 1998. NOVA Workforce Development based in Silicon Valley, is one of the most successful of these programs, partnering with companies to provide training and career services, and conducting research on local industry needs (Benner 2002: chap. 6).

Nonprofit/Community-Based Intermediaries

Although these programs typically receive the bulk of their funding from public sources, they are able to operate more nimbly and in more nuanced ways than is typical of public programs. For example, the Center for Employment Training (CET) is regarded as one of the more effective programs in the Valley. It has been successful in developing recruiting networks within the Hispanic community, as well as strong ties with networks of employers (Benner 2002: chap. 6).

Education-Based Intermediaries

Of course, all of the colleges and universities in Silicon Valley participate to a greater or lesser extent in efforts to connect their students to local employment opportunities. We describe some of these efforts in chapters five and six. We have already pointed to the importance of UC Berkeley Extension programs and Stanford's Honors Cooperative Program (see case example 3.A). These programs offer a range of training opportunities primarily targeting older, nondegree students who are seeking specific occupational skills. These programs are self-funding and administratively distinct from a mainstream University of California or Stanford education.

Governance Structures

Markets and Networks

No single body exercises fieldwide control over the San Francisco Bay Area. The firms and other types of organizations making up the region operate under a crazy quilt of governance bodies and regulations, including federal, state, county, municipal agencies, and special-purpose authorities. As noted, a handful of research institutes and policy associations, such as the Bay Area Council Economic Institute, Joint Venture Silicon Valley, and Silicon Valley Community Foundation provide research expertise, produce summary reports, and convene educational and policy-planning forums, but possess no formal authority. Other general regional agencies or associations also exist, such as the Association of Bay Area Governments, a regional planning group with representation from cities and counties, as well as specialized agencies such as Bay Conservation and Development Commission, which attempts to protect the ecology of San Francisco Bay.

In many ways, it would appear that the primary governance structure for the economic region of the Bay Area is the market. However, most informed observers of Silicon Valley insist that market forces within the region are tempered, if not subdued, by shared norms and strong interfirm and interpersonal ties (Saxenian 1996). As Seely-Brown and Duguid (2000: 16) point out, Saxenian importantly shifted "attention from economic issues to social and cultural ones," reminding us of Alfred Marshall's insight that "in localization social forces cooperate with economic ones." The economic structure of Silicon Valley has many of the features of a social network rather than a market or a hierarchy (a single integrated organization). As Powell (1990) points out, network systems blend aspects of cooperation and competition, relying on relational (ongoing) rather than short-term, arms-length ties. By stressing norms of reciprocity and open- ended, mutual benefits, they allow for greater flexibility and collaboration among the parties as they struggle (individually and collectively) to keep up with the frantic speed of change.

Metropolitan Media

While not precisely a governance system, community newspapers play an invaluable role in many communities, providing a common source of information, recording events of interest and import to local inhabitants, and oftentimes promoting a common identity and shared sense of community. Because the Bay Area is a large and complex metropolitan region, it is served by three major newspapers: the *San Francisco Chronicle*, serving primarily the North Bay region; the *Oakland Tribune*, serving the East Bay; and the San Jose *Mercury News*, serving the South Bay. Of these three, the *Mercury News*, with the subtitle "The newspaper of Silicon Valley," has attempted to serve as the primary megaphone of the Silicon Valley economy, chronicling the ups and downs of hi-tech industries and providing commentary on their trials and triumphs. The lead story in this newspaper is often devoted to the latest hiring spurt or successful IPO (initial public offering) or to controversies surrounding immigrant visas. Several columnists, from Michael Malone to Mike Cassidy and Michelle Quinn, have interpreted the arcane world of hi-tech to interested and affected laypeople. Such media play a large role in helping to develop a common discourse and hence structuring the regional field.

Assessing Human Capital

Some specialized oversight bodies are emerging in the intersection between higher education and the regional economy in the form of credential and certification authorities. This is a space that is of high interest to both colleges—who wish to contribute to the human and cultural capital of their students—and employers, who seek to acquire workers equipped with general knowledge and/or specific skills. To help ensure that employers can reliably obtain such workers, a complex array of professional, industry, and trade associations; commercial vendors; and governmental bodies has arisen to certify competence (Carnevale and Desrochers 2001).

Degrees, such as the BA (bachelor of arts), AA (associate of arts), and MS (master of science), are legitimated by accreditation bodies that evaluate colleges—their capacity, qualifications of faculty, compliance with rules and norms—and the satisfactory completion of specific course work by the student. As discussed in chapter two, they rely on measures of inputs rather than outcomes. The accreditation bodies are overseen by the Department of Education but regionally organized and constituted to represent the interests and uphold the standards of the professional academy. Degrees focus attention on the capacities and qualifications of educational organizations rather than the performance of individual students.

Certificates in many respects resemble degrees in that they also are based primarily on the evaluation of inputs by organizations: the satisfactory completion of specified programs. They differ from degrees in that they signify specialized education or training in a specific field, such as business management or health related fields. Certificates are awarded by public, nonprofit, and for-profit colleges and are often used to supplement degrees already obtained, signifying the acquisition of specialized knowledge. The number of certificates awarded by postsecondary institutions in the United States has grown substantially during the first decade of the twenty-first century, increasing from 2.3 to 3.4 million between 2000 and 2010. Public four-year institutions currently account for roughly 34 percent, public two-year colleges for 27 percent, nonprofit colleges for 17 percent, and for-profit schools 21 percent of certificates awarded. Of equal interest, in 2010 the four-year public institutions awarded slightly more certificates than degrees, two-year public colleges offered nearly twice as many certificates as degrees, nonprofit colleges awarded only about 10 percent certificates compared to degrees, four-year for-profits awarded slightly fewer certificates than degrees, whereas two-year institutions awarded five times as many certificates as degrees (National Center for Education Statistics 2013). Clearly, all colleges are increasingly engaged in awarding certificates, but two-year public and for-profit colleges are heavily involved in the business of offering certificates to students.

While the terminology is confusing, it is important to differentiate between certificates and certification. "*Certifications* are more often national, occasionally global, and always based on a standards-based assessment of student knowledge" (Carnevale and Desrochers 2001: 27; my emphasis). Rather than being based on inputs—for example, college curricula, faculty qualifications, or student "seat-time"—or on evaluations of the quality of organization providers, they rely on an assessment of individual student outcomes (learning, performance): external evaluations of knowledge and/or skills. Certifications are often established by national or international trade and professional associations who specify the criteria to be employed. Such organizations often contract with testing organizations, which conduct the assessments. A growing form of certification is created and conducted by "vendors," private companies such as Microsoft or Cisco, to attest that students are knowledgeable about a particular technology or product (Carnevale and Desrochers 2001; Patterson 1999). As is obvious, certification programs and supporting organizations have grown rapidly, but rather chaotically, in a free-market environment. Yet another indicator of accomplishment is provided by *digital badges*, which are in increasing use in regions such as Silicon Valley. They are designed to allow employees to claim and employers to recognize specific skills and knowledge acquired by the holder (see case example 3.C).

CASE EXAMPLE 3.C
DIGITAL BADGES

The concept of a digital badge originated in 2010 at a conference in Barcelona, Spain, held by Mozilla, the nonprofit organization that developed Firefox—a web browser whose software is "open source," free for anyone to download, copy, and improve (Ash 2012). Operating outside the conventional academic credit environment, digital badges are a means of validating discrete competencies online. Similar to the achievement-based patches connected to scouting organizations and video games, digital badges are essentially a web-enabled version of the traditional credential.

However, unlike a paper diploma or certificate, digital badges provide a granular picture of the credential, including qualifications of the educational provider, the date when the badge was issued, and evidence of completed work such as a portfolio of the students' assignments or their scores on a variety of tests and assessments (Carey 2015; Knight and Casilli 2012). Students can earn badges in a number of ways, such as passing standardized exams, watching videos, and contributing to class discussions.

"Badge issuers" are schools, employers, and other institutions that create a set of competencies or a curriculum as well as the assessments to determine if the "badge earner (i.e., the student)," has acquired the requisite skills for the badge. A number of government agencies (e.g., the National Oceanic and Atmospheric Administration), not-for-profit organizations (e.g., the Corporation for Public Broadcasting), businesses (Disney-Pixar), colleges, and universities are experimenting with digital badges—many of which complement traditionally earned credentials. The Agricultural Sustainability Institute at the University of California, Davis, offers a badge system that measures both formal (classroom) and informal (field internships) learning in sustainable agriculture and food systems (Fain 2014). In this competency-based program, learners hone their skills in systems-thinking, strategic management, and interpersonal communication, among other areas. In general, advocates (many of whom are in the technology industry) see badges as helping to bridge and legitimize educational experiences and skill development (e.g., "soft skills" such as teamwork) that happen both within and outside of formal school settings and throughout learners' lives.

Moreover, badges are "stackable" in that they can build upon one another and be combined with other badges across issuers. Such a structure allows learners to assess the skills that matter most to them and the options for gaining those skills, and gives the learner control in how they broadcast their qualifications to employers and professional networks (e.g., through social networking sites, blogs, personal websites, and career profile systems such as LinkedIn).

An online dashboard, also known as a "badge backpack," makes digital badges easier to share and verify, thus introducing a more transparent and information-rich system for checking the validity of credentials and qualifications, which had not been possible before (Acclaim 2014).

There are also questions and concerns related to badge issuers. Who should be able to develop, distribute, and award badges? Should these entities be certified or accredited in some way? At present, there is no independent accreditation system to evaluate online badge programs; as a result, it is possible that a badge from two institutions may or may not reflect the same level of skill.

More generally, the increasing use of certificates, certification, and badges signals a subtle but substantial shift from a reliance on traditional modes of recognizing and documenting educational achievement, such as degrees, which are authorized by academic institutions, to new ways of signaling competence that are endorsed by industry associations and employer groups. They represent a trend toward the increasing use of *employer-endorsed* modes of training.

A region like the Bay Area relies heavily on certificates, certification, and badges as an important basis for hiring new workers for many of the lower- and mid-level positions to be filled. Because of their specificity and external validation, would-be employees stress their acquisition and employers give primacy to such signals of competence in their web-based search for workers. However, for many kinds of positions, companies recognize that specific skills are no substitute for general kinds of knowledge of the kind associated with educational degrees. No less an icon than Steve Jobs, founder of Apple, pointed out that "it is in Apple's DNA that technology alone is not enough. It's technology married with liberal arts, married with the humanities, that yields us the result that makes our hearts sing" (Carmody 2011). Students with degrees are still in demand by many companies and for many positions.

Higher Education and the Regional Economy: Tensions between Two Fields

As we have attempted to demonstrate, both higher education and regional economies are highly complex organizational fields: each hosts a varied and constantly changing ecology of organizational forms and their members. Both are supported and constrained by numerous players outside of the geographical boundaries. The fields are somewhat independent, but also highly interdependent. We begin by pointing out two important features that are common to organizations operating in the two arenas.

First, both colleges and central players in the Bay Area hi-tech economy share a common interest in what is often termed a "knowledge economy." High value is placed on expertise, information, and knowledge, although the academic players place a premium on abstract knowledge while the hi-tech firms emphasize applied knowledge and know-how. Second, both types of fields, particularly at the upper levels, create dense interactional networks in which value is placed upon expertise and a reputation for reliable performance. Both universities and hi-tech firms benefit from strong network ties that connect, in the former case, faculty and administrators and, in the latter, companies and their leadership groups into wider communities within which competitors can become potential partners. Knowledge and information are shared not only within organizations but also across them, fostering common norms and values. In these conditions, competitive firms in multiple industries, ranging from semiconductors, legal expertise, and design to marketing and venture capital, can develop "reputations in the Valley [that] replace the need for detailed legal contracts and whose reputations, earned through performance with many different customers, make them easy to trust at a technical level" (Seely-Brown and Duguid 2000: xv).

We note, conversely, several important ways in which the fields of higher education and hi-tech regions differ (table 3.1). First, organizations within each field differ in the primary types of *control mechanisms* they confront. Higher education is subject to regulative controls enforced by public agencies, and normative controls enforced by academic and professional associations. Colleges, especially public ones, experience multiple levels of control: federal, state, regional, and local (district or municipal). Regulative controls emphasize conformity to a wide range of rules regarding curricula, qualifications of faculty, accountability for public funds, and attention to the rights of students. Normative controls stress adherence to academic and professional standards, enforced by multiple associations and accreditation bodies. Most of these controls concern either *structural features* that colleges must possess (e.g., libraries and other learning resources, appropriate governance and administrated structures, faculty qualifications) or *processes* that must be followed (e.g., state approvals, alignment of programs and missions, accounting procedures). In short, process and input controls are relied upon heavily. Measures of *outcomes* (e.g., graduation rates or evidence of student learning) have up to now been much less widely employed (see chapter two).

Table 3.1   Contrasting Structures and Logics of Colleges and Silicon Valley Firms

	Higher education	Silicon Valley
Governance systems	Regulative (federal, state) Professional norms Process controls	Markets Networks Outcome controls
Institutional logics	Liberal arts Learning skills Theoretical knowledge Preservation	Practical arts Vocational skills Applied knowledge Disruption
Prime beneficiary	Public good	Private interests
Stage of development	Mature	Early, formative
Time horizon	Centuries; decades	Years; months
Pace of change	Deliberate	Rapid

In contrast, firms and other organizations within the regional economy, while subject to various federal and state tax requirements, environmental regulations, and rules concerning labor and consumer relations, are primarily governed by a variety of *outcome* metrics. These market controls take a variety of forms, including simple measures of output produced, sales, or growth, but most involve ratios assessing value added such as productivity or return-on-investment (ROI) metrics. For publicly traded companies, the latter ROI-type measures, often assessed on a quarterly basis, are highly salient to financial markets and their investors.

The two fields are governed by somewhat contrasting *institutional logics* (see Gumport 2000). As previously defined (chapter two), institutional logics are sets of "material practices and symbolic constructions which constitute [a field's] organizing principles and which are available to organizations and individuals to elaborate" (Friedland and Alford 1991: 248). As noted, players in both fields value knowledge, expertise, and creativity, but with different emphases. All colleges have been shaped and are influenced by the historical legacy of academic traditions spanning over several centuries. Colleges and universities view themselves as the custodians and curators of the vast knowledge and cultural heritage of the past: the philosophical and scientific constructions, the great languages, and literatures. They have also long emphasized their contribution to the public good: preparing informed and thoughtful citizens and members of society. Scientific studies are undertaken to broaden our understanding of how our world operates and to expand the frontiers of knowledge, not primarily because of their applications to specific problems or to secure short-term benefits. There is respect for tradition for its own sake, as symbolized in the elaborate rituals of graduation and investiture. In many respects, colleges are more like "temples" than businesses (Meyer 1977; Stevens, Armstrong, and Arum 2008). This academic logic, as noted elsewhere, was never as strong among colleges in this country as in Europe, and it varies greatly among the different types of colleges. Still, there remains a strong core belief that colleges and universities are the keepers and protectors of a precarious value: the preservation and enhancement of the stock of common knowledge, the wisdom of the ages. Students in colleges are more likely to be regarded as "clients" whose needs should be diagnosed and met, not as "customers" whose preferences are to be served.

Private firms in the regional economy are, of course, much more oriented to market logics. They are guided by the changing tastes and demands of their consumers. They turn to the colleges and universities for help in solving their specific problems and, in particular, to obtain a qualified workforce trained to contribute to their research, development, and production systems. They value creativity and knowledge insofar as it leads to commercially successful products and services. Students and employers view education primarily as a private good: specific skills and knowledge possessed, to varying degrees, by individuals.

The frictions engendered by these varying logics are well illustrated by the efforts of Alan G. Merten, president of George Mason University in Virginia in the late 1990s, to provide a rationale for the University. He stated, "We have a commitment to produce people who are employable in today's technology workforce." He argued that their students are "good consumers" who want degrees in areas where there are robust job opportunities, and the university has an obligation to cater to that demand. In response to George Mason's cuts in the budgets of the humanities departments, 180 professors sent a letter to Merten in 1998 acknowledging that while training students for the job market was a legitimate goal, "precisely in the face of such an emphasis on jobs and technology, it is more necessary than ever to educate students beyond technological proficiency. . . . The central mission of George Mason University should be to educate self-reflective, knowledgeable, critically thinking and responsible citizens upon whom a humane society depends" (quoted in Washburn 2005: 214).

Finally, as implied in the foregoing discussion, the sector of higher education and the field of the regional economy operate in different time frames. Higher education is engaged in the pursuit of long-range and enduring truths. It views itself as part of a centuries-long effort to accumulate and test knowledge and to cultivate the humanities and arts. It also has the responsibility to pass on this heritage by educating and socializing students. Education takes time. The "production schedules" of colleges and universities are measured in years—four for a BA, more for an MA or PhD—and in semesters or quarters. Colleges set and are guided by long-range goals: they pursue "missions" and are funded by "patient" money: philanthropists, foundations, and governments. These types of funders are not expecting a quick return on their investments but, rather, are supporting the long-term accumulation of knowledge and the advancement of distant, long-range goals (e.g., a cure for cancer or improved understanding of global interdependencies).

In contrast, firms and other players in hi-tech economies are subject to strong competitive forces to be first to market with an improved product or service. Their production systems are expected to rapidly respond to changes in market signals. Both companies and their employees live in a world of high energy, uncertainty, and transience. It is a world of flux in which industry and firm boundaries are reconfigured, workers redistributed, and new and different alliances are forged. Firms are primarily valued by their facility for their rapid response to new and different challenges (Lewis 2000).

The tensions we have depicted refer to the modal college and company. There exists a wide distribution of colleges, ranging from those striving to be academically pure to those willingly embracing the values of the marketplace. Moreover, the components or subunits of colleges vary in their orientation: some may focus on long-term scholarship, others on workplace demands. In a similar way, firms vary greatly in how much pressure they experience from market forces and how rapidly demands and technologies change. Still, the controls, logics, and time scale vary enough that we must ask: how is it possible for organizations within these two arenas to connect and collaborate? We explore this broad question here and in the following chapters.

Adaptation by College Populations to the San Francisco Bay Area Economy

How well have colleges in the San Francisco Bay Area adapted to the rapidly changing labor demands of the Silicon Valley economy? In addressing this question, we need to bear in mind that any changes we observe are partly responsive to broader developments occurring in the US economy and in higher education generally during the past four decades. We consider some of these general changes in chapter four. Still, there seems little doubt that colleges do experience in a very direct way the pressures brought to bear on them to respond to the needs of the regions in which they reside. Here we report quantitative data on major ways in which our populations of degree-granting colleges have dealt with the challenges posed by the economy in which they are embedded. Subsequent chapters will provide a detailed and nuanced answer to this question, focusing more on *how* colleges have made changes—the specific strategies employed—as well as attending to differences among sub-regions in demands made and opportunities presented.

We have stressed that one of the principal features of the Silicon Valley economy is the pace of change: the rate at which new types of firms and jobs are being created that demand new types of skills from workers. Also, we have pointed out that most colleges are not designed to be quickly responsive to changes in the types of programs they provide. One of the hallmarks of traditional colleges is the faculty tenure system, a normative-sanctioned employment structure designed not only to support freedom of thought and inquiry for individual faculty members, but also to ensure that colleges provide continuity in curriculum and academic practices. However, during the past half-century, US colleges have witnessed a steady decline in the ratio of tenured faculty employed (see Clawson 2009). This change is observed in the Bay Area. Table 3.2 reports data by type of college in the Bay Area on the number of tenure and nontenured faculty members between 1995 and 2010. In every college type, the ratio of nontenure-track to tenure-track faculty has increased since 1995. But, as expected, there are important differences among types of colleges. For instance, in public four-year systems, the number of tenured faculty has remained relatively constant, but the number of nontenured teachers has increased from 31,018 in 1995 to 38,064 in 2010. By contrast, in for-profit colleges, tenure-track faculty are virtually nonexistent in both two- and four-year for-profit colleges. Faculty in these proprietary institutions are hired and fired to meet the demands of a changing market.

The significance of these data is not simply that they indicate important changes in the internal structure of colleges. Nontenure-track faculty differ from tenure-track faculty in two important respects: first, they provide colleges with much greater flexibility. Most nontenured faculty are hired as adjunct professors, part-time and short-term employees. They allow colleges to better cope with demands posed by rapid changes in the numbers and interests

Table 3.2   Number of Tenure-Track and Nontenure-Track Staff, by Institutional Sector, San Francisco Bay Area Colleges, 1995–2010

	Public 4-year		Nonprofit 4-year		For-Profit 4-year	
Year	Tenure teaching	Nontenure teaching	Tenure teaching	Nontenure teaching	Tenure teaching	Nontenure teaching
1995	3,863	31,018	2,211	13,844	0	782
2000	2,994	32,113	2,089	13,473	0	1,447
2005	2,665	25,519	1,845	16,220	0	2,424
2010	3,929	38,064	1,892	13,028	0	2,896
	Public 2-year		Nonprofit 2-year		For-Profit 2-year	
Year	Tenure teaching	Nontenure teaching	Tenure teaching	Nontenure teaching	Tenure teaching	Nontenure teaching
1995	2,826	10,798	9	427	2	307
2000	3,398	10,456	0	340	4	415
2005	3,663	11,077	0	164	0	530
2010	3,328	9,936	0	0	0	635

Source: Integrated Postsecondary Education Data System.

Note: Tenure considers personnel in positions that lead to consideration for tenure. We define nontenure personnel as the difference between total staff and tenure-track staff.

of students. Second, adjunct faculty are much more likely to teach applied or vocational than academic courses. An increase in their numbers signifies a likely change in the types of degrees and courses offered to favor vocational/ occupational training over general education.

To directly examine this shift in types of academic programs, figure 3.2 reports changes in the number of degrees conferred by type of college in the Bay Area between 1974 and 2010. The number of degrees granted by all types of colleges reflects a substantial shift from liberal arts programs, such as the humanities and natural and social sciences, in favor of vocational and applied, professional and paraprofessional programs. In public four-year programs, liberal arts degrees conferred increased somewhat from 20,000 in 1974 to roughly 25,000 in 2010. In nonprofit four-year colleges, they also increased slightly during the same period, from just under to just over 5,000. By contrast, during the same period, occupational programs nearly doubled in liberal arts public colleges from 16,000 to 30,000. And in nonprofit programs, the rate of increase was even higher, from 6,000 to 14,000. Clearly, one of the survival strategies for nonprofit four-year colleges has been to increase their offerings in vocational and professional training, although there are important exceptions,

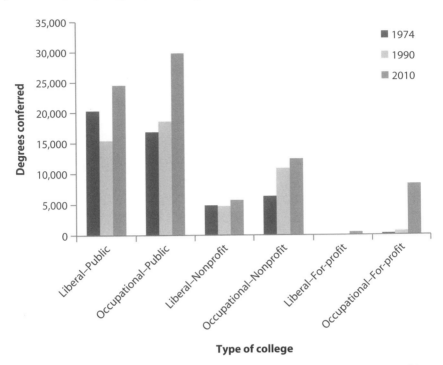

Figure 3.2. Field of degrees conferred: Liberal arts and sciences and occupational/professional, San Francisco Bay Area Colleges, 1974–2010.
Source: Higher Education General Information Survey (1970–1985) and Integrated Postsecondary Education Data System (1986–2012).

such as Mills College. Again, for-profit colleges, both two- and four-year, present a completely different profile. These programs granted very few liberal arts degrees in 1974 but had begun to offer some of these degrees by 2010. Their significant increases have been almost entirely in occupational degrees. This is the terrain on which for-profits have made their advances.

During the period of our study, all types of colleges have moved to embrace a vocationally oriented curriculum, as proportionally fewer students elect to pursue a liberal arts education. Public four-year colleges, as expected, have been slowest to embrace these changes, but both nonprofit four-year colleges and public community colleges have moved substantially in this direction. Understandably, for-profit schools are largely devoted to vocational training. As noted, they are hierarchical and centralized, and their managerial systems are designed to assess and take advantage of changing market demands, replacing faculty to quickly respond to developing training opportunities.

The general shift away from the liberal and toward the vocational applies primarily to four-year degree programs. The story changes in an examination of career and technical education (CTE), a category that covers both nondegree certificate programs and vocational associate's degrees that are not designed to transfer into baccalaureate institutions. CTE programs in California are concentrated at community colleges, and students are gravitating away from them. Between 2001 and 2014, the percentage of full-time equivalent (FTE) enrollments in CTE courses fell from 31.3 percent to 28.2 percent (Blue Sky Consulting Group 2015). Furthermore, enrollments in CTE are highly concentrated; while a recent study identified 142 CTE fields of study offered at California community colleges, over half of all enrollments were in just eight of those fields. Statewide, the most popular field is administration of justice, which is the second most popular field in the Bay Area's colleges, closely following nursing (Moore et al. 2012).

CTE programs face other challenges besides diminishing, and concentrated, enrollments. Historically, program completion has been extremely low. In the cohort of over 255,000 students that entered California community colleges in 2003–4, only 5 percent had earned CTE certificates within six years, and only 3 percent had earned CTE associate degrees (Shulock, Moore, and Offenstein 2011). Additionally, CTE course work is significantly more expensive to provide, due to both higher start-up and operating costs. This is not only true for highly specialized fields; in 2011–12, instructional cost per student hour was $131 for allied health courses, compared to $64 for general biology courses (Shulock, Lewis, and Tan 2013). This means that CTE course work is often an easy target for cuts in lean budget years (Blue Sky Consulting Group 2015). Finally, our program review data, discussed in chapter five, indicates that despite strong commitments to CTE by institutional leaders, countervailing preferences among both students and instructors boost the desirability and availability of traditional arts and sciences classes, especially those that are transferrable to four-year institutions.

Chapters five and six provide a detailed discussion of the ways in which the various types of colleges have coped with the changing demands of their environments. While such changes are particularly pronounced in a rapidly changing economic region such as the Bay Area, similar trends are occurring across the country, as Brint (2002) documents in his study of all four-year, degree-granting colleges for the entire United States during the period 1970 and 1995.

Concluding Comment

The San Francisco Bay Area region is remarkable in a number of respects. It represents a leading example of the "new" economy as it began to emerge in the latter decades of the twentieth century. It is an economy strongly dependent on knowledge, expertise, and innovation and, as a consequence, relies heavily on the involvement of higher education centers as the basis for its on-going success. Not only research universities but also state universities, community colleges, and assorted nonprofit and for-profit colleges are consequential players in this type of region. The economy is made up of multiple types of organizations—firms, venture capitalists, law firms, colleges, and various intermediaries that broker relations among them. The labor market is notable for its flexibility and volatility. While competitive markets exercise strong controls, the new economy also is supported by shared norms and shifting relational networks among individuals and organizations.

The fields of higher education and the regional economy intersect and overlap, but they exhibit substantial differences and experience resulting tensions. Colleges stress academic values, the preservation of existing knowledge, and do not accommodate change easily.

The Silicon Valley economy is subject to rapid change, both in terms of what types of industries dominate and what kinds of knowledge and skills are required. Conventional colleges have attempted to meet some of these demands by devoting more attention and resources to new courses of study and vocational training but find it difficult to keep pace with Valley needs. It appears that some of this demand is met by the increasing use of corporate colleges and internal training programs as well as by large numbers of short-term, specialized schools that have blossomed in recent years in the postsecondary education sector, but data on these programs are elusive and often proprietary. Colleges have explored a variety of strategies in their struggle to keep pace and partner with firms in this region. These efforts are reviewed in subsequent chapters.

4

Broader Forces Shaping the Fields of Higher Education and the Regional Economy

W. RICHARD SCOTT, MANUELITO BIAG,
BERNARDO LARA, AND JUDY C. LIANG

In chapters two and three, we described the two organization fields that are central to our interest: the field of higher education in the United States, and the regional field of the San Francisco Bay Area economy. As we have emphasized, both fields are made up of a complex ecology of diverse, interacting organizations, and both have undergone substantial change during the past 40 years. In subsequent chapters, we explore in more detail how these two fields have engaged with one another and how, over time, they have become closely interconnected. We also observe some of the results of these encounters for colleges of varying types—in altered strategies, in new programs and procedures, and in reconfigured organizational structures.

Before we pursue further the task of connecting the two fields, we need to attend to broader societal forces that have been at work in shaping these fields throughout the period of our study. All organization fields are, by definition, subsystems of broader systems, and all fields are affected by developments occurring outside their functionally delimited boundaries (Friedland and Alford 1991). It is not possible to depict all of the broader forces at work, but we believe that five are particularly salient:

1. Demographic changes occurring at the state and regional level
2. Lifestyle changes occurring at the societal level
3. Organizational changes occurring at the societal level
4. Political and economic forces shaping higher education, including national and state structures and policies affecting both colleges and workforce training
5. Technological changes, especially changes in information and communication technologies.

Broader environments influence specific fields, but the reverse is no less true: fields can exert effects on their environments. A region such as Silicon Valley cannot be viewed as a passive pawn; it plays a strong role in shaping its demographic, economic, political, and technological environment. As one of the most innovative and fastest-growing metropolitan regions in the world, the Bay Area strongly shapes its broader context.

Demographic Changes

California has sustained more population growth year after year than any other developed area of the same size. The state's population nearly tripled in the last half of the twentieth century, growing to roughly 39 million residents in 2010. During the period 1970–2010, the San Francisco Bay Area nearly doubled in size, growing more rapidly than the rest of California, and much more than the nation as a whole. While California's higher education sector has attempted to keep pace with its growing population by rapidly expanding the number and types of colleges (table 4.1), the state has failed to expand research and state universities as rapidly as necessary to meet increasing demands.

Immigrants have long been a major force in helping California grow to be the seventh-largest economy in the world. Large groups of immigrants from more than 60 countries call California home. Estimates from the 2012 American Community Survey indicate that about 53 percent of immigrants into California came from Latin American countries, particularly Mexico, and

Table 4.1   Population Growth in California and Higher Education Enrollment, 1960–2005

Year	California population (thousands)	Population growth (%)	Total growth in public higher education enrollment (%)
1960	15,727	49*	67**
1970	20,038	27	300
1980	23,780	19	36
1990	29,828	25	12
2000	34,099	14	16
2005	36,154	6	14

Source: Callan 2009.
Note: Population and enrollment growth figures are for the previous decade, except for 2005 figures, which are compared with 2000.
*Increases are for decade ending in 1960.
**Enrollment data are for fall full-time-equivalent students.

37 percent from Asian/Pacific countries, especially China, India, and the Philippines. Compared to the rest of the country, California has a much lower share of its immigrants coming from Europe or the Caribbean. Most of its growth in recent decades has been the result of increased immigration from Asian and Latin American countries.

As with the state as a whole, the ethnic composition of the Bay Area has changed dramatically in recent decades: between 1980 and 2010, white residents declined from 3.6 to 3.0 million while numbers for African-Americans Americans remained largely unchanged at .5 million (table 4.2). Meanwhile, Hispanic and Asian/Pacific populations increased substantially, from .6 to 1.7 million for Hispanic residents and from .4 to 1.7 million for Asians. While increases in these flows are roughly the same magnitude, their composition is quite different: Hispanics arrive with lower levels of education and exhibit lower levels of occupational skills than do Asian immigrants.

Clearly, Silicon Valley has been a magnet attracting large numbers of Asian immigrants to the Bay Area. Saxenian (2006: 53) documents the differences between immigrant groups in populating the professional/technical occupations of Silicon Valley (fig. 4.1). During the period 1985–2000, workers of Chinese and Indian origin together accounted for more than 57 percent of professional jobs in the tech sector, while immigrants from Mexico accounted for only about 3 percent. The numbers are even larger if one counts the flow of Asian and Indian "nonimmigrants" admitted for periods of up to six years with H-1B visas (see chapter three). Recent data report that an astonishing 58 percent of Silicon Valley residents employed as STEM professionals are foreign-born, substantially more than in comparable hi-tech regions (table 4.3).

Table 4.2 Population Growth in the San Francisco Bay Area, by Race/Ethnicity, 1980–2010

Year	Total	White (%)	African American (%)	Asian (%)	Hispanic (any race) (%)
1970	4,628,199	86.4	7.9	4.4	8.2
1980	5,179,784	69.6	8.9	8.4	12.2
1990	6,023,577	60.7	8.6	14.7	15.3
2000	6,783,760	50.0	7.3	19.3	19.4
2010	7,150,739	42.4	6.4	23.6	23.5

Source: US Census.
Note: Census data from 1970 to 1980 count Hispanics as "Persons of Spanish Origin or Descent." Data from the 1970 census count African Americans as "Negroes."

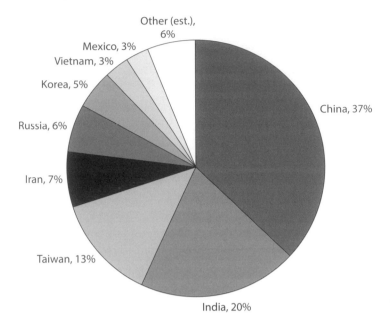

Figure 4.1. Professional and technical immigration to the San Francisco Bay Area, 1985–2000.
Source: Saxenian 2006.

Large numbers of workers from other areas such as Mexico and Latin America have been lured by the vibrant Bay Area economy, but most of these immigrants occupy lower-level technical and support positions or work in other sectors, such as health services (see case example 4.A).

Table 4.3   Foreign-Born and In-State-Born Share of Population in STEM Professions, with a Bachelor's Degree or Higher, 2014

	Foreign-born share (%)	In-state-born share (%)
Silicon Valley	58	20
New York City	43	27
Southern California	39	34
Seattle	34	21
Boston	32	31
Austin	25	36

Data Source: US Census Bureau, PUMS.
Note: Used with permission from the Silicon Valley Competitiveness and Innovation Project.

CASE EXAMPLE 4.A

IMMIGRATION

Contributions of Immigrants in Silicon Valley

Immigrants have played an essential role in the development and success of the Bay Area's technology economy. Roughly 25 percent of Silicon Valley's workers are foreign born, often employed in highly skilled positions such as engineers, scientists, and senior managers in software, semiconductor, and internet content and services (Saxenian 2002a).

Given the significant contributions that foreign-born workers have made in Silicon Valley, and with tech companies such as Facebook concerned about a shortage of skilled talent, technology leaders and politicians alike have supported immigration reform bills to increase the flow of educated foreign-born workers. These bills include the Startup Act, which would create a visa category for foreign entrepreneurs as well as the Immigration Innovation Act, which would increase the cap on H-1B Visas from 65,000 to 115,000 (Benner 2015).

The immigration trajectories of foreign-born workers in Silicon Valley are varied. Many foreign-born engineers and scientists are products of US graduate programs. In 2003, foreign-born students earned 40 percent of US doctoral degrees in science and engineering; they were responsible for a majority of the growth in doctoral degrees in these fields between 1985 and 2005. Students from four countries—China, India, South Korea, and Taiwan—accounted for one-half of the doctorates awarded to foreign-born students during that period (National Science Board 2010). The same report also estimates that approximately two-thirds of foreign PhD students stay in the United States at least five years after earning a degree from an American university. In contrast, other immigrants came to the United States after being recruited by intermediary organizations (Saxenian, Motoyama, and Quan 2002).

The Role of Immigrant Associations

Regardless of their pathways, many immigrants have formed strong social networks within the Bay Area, often becoming members of associations that are religious, gender-based, professional, or ethnic-based in nature. Among others, these associations include the Silicon Valley Chinese Engineers Association, Women in Technology International, and the Indus Entrepreneurs. These groups are critical to the success of many immigrants, as well as minority groups, because they help foster social networks, encourage information exchange, and provide support structures for new ventures, thereby advancing workers' careers and prospects (Saxenian, Motoyama, and Quan 2002). These networks also provide successful and established role models for aspiring entrepreneurs, help connect younger employees with seasoned professionals, serve as recruitment channels to various job opportunities, provide key resources and trainings (e.g., English communication, negotiation skills, how to create a business plan), and help establish partnerships with US-based companies (Saxenian 2002b).

In addition, many of these associations help workers connect with other business partners in their home countries. Despite coming to the United States, many of these highly educated immigrants have transnational entrepreneurial ambitions, and with cultural, language, and technical know-how, most hope to build economic and social bridges between the United States and their native countries. The Silicon Valley model of industry, which is characterized as a "fragmented industrial structure organized around networks of increasingly specialized producers" (Saxenian, 2002a: 2–3), encourages these types of global ambitions.

"The Invisible Workforce"

Despite the well-known success of highly educated foreign-born workers in Silicon Valley's hi-tech industries, a closer examination of the region's immigrant workforce yields glaring disparities. One in five low-wage contract service workers in Silicon Valley are immigrants; these include groundskeepers, janitors, security guards, cooks, and shuttle drivers, among others. While these workers have contributed substantially to the success of the region, a majority struggle to make ends meet, often making less than $16 an hour (compared to $63 an hour for software developers), which is not enough to meet the basic standard for family self-sufficiency in the region (Auerhahn et al. 2012). As contract workers, many also lack access to critical benefits, including paid sick leave. The wage gap between less educated and highly educated workers in the Bay Area continues to grow. Given the challenges of living in a high-priced region like the Bay Area, many of these low-wage immigrant workers live at or near the poverty level; as such, many have organized to pressure hi-tech companies such as Intel, Facebook, Google, Genentech, Apple, and Yahoo for higher, livable wages and access to benefits (Wilson 2015b).

Implications for Higher Education

Foreign-born students continue to make up an increasing share of the student body in many postsecondary institutions. For example, the number of students on F-1 visas in US colleges and universities grew from 110,000 in 2001 to 524,000 in 2012, with sharp increases from countries with emerging economies such as Saudi Arabia and China (Ruiz 2014). About two-thirds of these students are pursuing education and training in STEM or business, marketing, and management fields. Moreover, recent data from the Labor Department suggest that among full-time wage and salary workers in the United States with a bachelor's degree or higher, foreign-born workers earned, on average, higher wages than their native-born counterparts. While median wages increased from 2010 to 2014 at about 4 percent for native-born college graduates, they rose 9 percent for foreign-born workers with an undergraduate degree or higher (US Department of Labor 2015).

Still, generalizations about the educational attainment of immigrant populations can be misleading since outcomes among different groups are so diverse. While Chinese, Indians, and Koreans have among the highest levels of education, immigrants from Mexico, El Salvador, Laos, and Cambodia have among the lowest. Silicon Valley's level of educational attainment is uniquely high, with about 46 percent of adults having a bachelor's degree or higher, compared to 31 percent of California adults and 29 percent nationally (Massaro and Najera 2014). Although the proportion of Asian adults with at least a bachelor's degree has risen, the share of Latino adults with a college education has remained stagnant. In fact, according to the National Center for Education Statistics, while college enrollments have risen significantly since 2000, degree completions have not kept pace, particularly among Latino first-generation college-going populations.

With workforce development occurring primarily in broad-access colleges, significant reforms and investments will be necessary to ensure that these institutions can effectively prepare workers for future jobs, particularly in newer middle-wage jobs, in advanced manufacturing, information, and health technology (Holzer 2015). Community colleges in California lack sufficient resources to offer students high-quality course options, academic and career counseling, and effective remediation. Many broad-access colleges also lack the capacity as well as the incentives to respond to local labor market needs. High-demand classes such as those in information technology or health-related fields quickly become oversubscribed. Moreover, some institutions focus less on helping students transition to the workforce and more on feeding students into four-year colleges (Backes and Velez 2015). Under these circumstances, for-profit colleges are able to fill a niche since they face stronger incentives to respond to labor market needs (Deming, Goldin, and Katz 2013).

Immigrants arriving with training are less dependent on regional educational programs but, in general, population growth in the Bay Area has placed strong demands on its higher education system. Not only have the increased numbers posed challenges, but colleges and universities have been confronted with novel needs presented by "nontraditional" students, including those from different backgrounds, academic preparation, expectations, and languages. Figure 4.2 depicts the share of young people from diverse ethnic groups, ages 15–24—the cohort more likely to demand educational services—as it has changed between 1990 and 2010. Note that while white numbers have been in decline, Hispanics have increased to the point that by 2010 their numbers are almost equivalent to whites in this age group. Asian/Pacific 15–24 year olds have also increased in numbers, but at a slower rate than Hispanics. This trend suggests

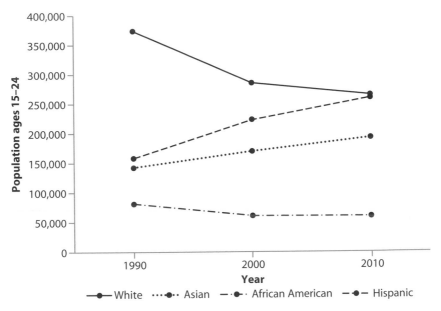

Figure 4.2. Population ages 15–24 in the San Francisco Bay Area, 1990–2010.
Source: US Census.

that Hispanics may be exerting greater pressure on Bay Area higher education than their Asian/Pacific counterparts. The number of African Americans in this age bracket has declined somewhat during this same period.

Table 4.4 reports the enrollment composition as it is distributed across the various types of colleges in the Bay Area during the period 1982 to 2010. Gratifyingly, it appears that college enrollments broadly reflect the demographic changes occurring in the region during this period. At the same time, differences in enrollments across colleges are revealing. White students are overrepresented in the selective private nonprofit colleges, whereas Hispanic students are underrepresented in four-year public colleges and overrepresented in the public two-year community colleges. Asian students appear to be somewhat underrepresented in private colleges, both for-profits and nonprofits. African American students are underrepresented in public four-year programs but overrepresented in private

Table 4.4   Percent Enrollment by Race/Ethnicity across Types of Colleges, San Francisco Bay Area, 1982–2010

	Year	White	African American	Hispanic	Asian	Other
Public 4-year	1982	0.66	0.06	0.06	0.16	0.06
	1990	0.55	0.06	0.10	0.24	0.06
	2000	0.39	0.06	0.13	0.34	0.08
	2010	0.35	0.05	0.18	0.31	0.11
Public 2 year	1982	0.69	0.09	0.07	0.12	0.03
	1990	0.59	0.08	0.11	0.17	0.04
	2000	0.41	0.09	0.17	0.30	0.03
	2010	0.32	0.09	0.25	0.28	0.06
Private nonprofit	1982	0.73	0.04	0.04	0.07	0.11
	1990	0.68	0.05	0.06	0.11	0.11
	2000	0.54	0.06	0.10	0.17	0.12
	2010	0.46	0.06	0.13	0.17	0.18
Private for-profit	1982	0.47	0.09	0.10	0.25	0.09
	1990	0.54	0.10	0.18	0.11	0.07
	2000	0.41	0.09	0.15	0.20	0.15
	2010	0.33	0.13	0.21	0.16	0.17
Total Bay Area	1982	0.68	0.08	0.07	0.12	0.05
	1990	0.59	0.07	0.10	0.18	0.05
	2000	0.43	0.08	0.15	0.29	0.06
	2010	0.35	0.08	0.21	0.27	0.10

Source: IPEDS.

for-profit colleges. More generally, Hispanic students appear to be taking advantage of low-cost community colleges, whereas African American students are more likely to opt for higher-priced for-profit enterprises.

Part of the explanation for differences in ethnic distributions of students across types of colleges is revealed in the differences among ethnic groups in the qualifications of high school graduates, as reported in table 4.5. While data are only available beginning in 1993, they reveal substantial differences between the graduates of the four major ethnic groups in the percent that have completed all courses required for entry to the two top tier systems: CSU and UC. Asian students exhibited the highest rate for this indicator of achievement, followed in order by white Hispanic, and African American students. However, note also that the percentage of qualified students increased for every ethnic category between 1993 and 2014. When combined with the overall increases in population during this period, as reported in table 4.1, this

Table 4.5   High School Graduates Completing All Courses Required for Entrance to the University of California and/or California State University System, by Race/Ethnicity, 1993–94 to 2013–14

	1993–94 (%)	2000–2001 (%)	2005–6 (%)	2010–11 (%)	2013–14 (%)
African American	28	25	24	24	31
	(3,639)*	(3,785)	(3,868)	(4,084)	(4,013)
Asian	52	58	60	63	69
	(10,715)	(13,041)	(14,904)	(15,873)	(17,153)
Hispanic	18	23	23	27	35
	(7,419)	(9,580)	(10,769)	(15,752)	(17,840)
White	40	48	51	54	61
	(18,960)	(20,665)	(20,423)	(18,446)	(17,401)
Total	38	44	45	46	53
	(40,970)	(47,540)	(51,441)	(56,415)	(58,716)

Source: DataQuest, California Department of Education.
* Numbers in parentheses represent the number of graduates.

Note: "Completing UC/CSU Required Courses" includes the number of twelfth-grade graduates, for the school year indicated, completing all the courses required for University of California (UC) and/or California State University (CSU) entrance with a grade of "C" or better. This represents only a portion of the entrance requirements for UC or CSU. Data in this table include the following counties: Alameda, Contra Costa, Marin, San Francisco, San Mateo, Santa Clara, Santa Cruz. The "Asian" category includes those students who identified as "Asian," "Filipino," or "Pacific Islander." Students who identified as "American Indian or Alaska Native" or "Two or more races" or did not identify race/ethnicity are not reported as subgroups; however, these groups are included in the "Total" values.

change indicates why California's top colleges confront enormous challenges: the number of qualified students for CSU and UC has expanded more rapidly than the number of spaces to accommodate them, as noted in chapter two (see fig. 2.4). We elaborate on this problem in this and subsequent chapters.

Lifestyle Changes

In addition to servicing an ethnically diverse collection of students, colleges also have to deal with students that vary in their aspirations and their location in the life course (Settersten 2015). The period beginning in the 1960s witnessed a large increase in the proportion of high school students with aspirations to seek a college education. Today about two-thirds of high school students want to graduate from a four-year college and one-fourth from a two-year college (Aud et al. 2012). As a consequence, the demand for access to a college education is at an all-time high in this country. And, as a corollary, the current pool of applicants exhibits much larger diversity than previous cohorts. They are more likely to be older, married, in need of remedial courses, to attend part time, and live off-campus. For example, in 2008, 53 percent of first-year US undergraduates attended part time, 38 percent were over 24 years of age, 36 percent were in remedial courses, and 25 percent had financial dependents. Only 13 percent of undergraduate students lived on campus (National Center for Education Statistics 2010: table 240). In short, colleges of all types, but especially broad-access colleges—are expected to adapt to meet the needs of many types of nontraditional students (Deil-Amen 2015).

Even these data understate the changes in the types of students seeking to take college courses. Colleges are expected to serve not only high school graduates but also older students seeking additional educational preparation, including new types of skills. Adult education has long been a part of the portfolio of broad-access colleges, but traditionally these were older adults seeking personal enrichment or to develop an avocation. Many postsecondary training programs have also catered to the needs of working adults who wish to acquire or enhance their job skills. However, with the advent of new industries characterized by rapidly changing technologies, an increasing number of rank-and-file employees are either dislodged from their current jobs and obliged to seek new skills or return to college to acquire education and training to improve their job chances. Colleges are expected to provide not only training for would-be workers but also retraining for the current labor force.

To capture and assess the magnitude of these changes, a recent survey of 3,200 current and prospective undergraduate students conducted by the Parthenon Group attempted to estimate the size of student segments in today's colleges. They report that roughly 35 percent of these students may be characterized as having focused academic interests or an exploratory orientation that includes enjoyment of team sports and a social life. About 18 percent are less interested in academic degrees than in vocational training. However, a large proportion of students (21 percent) are either "career accelerators," who are currently employed and attending college part time to enhance their job skills, or "industry switchers" (18 percent), who are unemployed or in a low-growth industry and seeking skills to change their careers. Finally, 8 percent were considered to be "academic wanderers," who lack coherent educational goals or plans (parthenon .ey.com/po/en/perspectives/the-differentiated-university-introduction).

The demands placed on colleges by the increasing diversity of their students are much greater today than at any time in the past.

Organizational Changes
Differentiation

California is not alone in experiencing population increases. If we consider US society as a whole, the population has increased from about 4 million residents in 1790 to over 300 million today. The fastest rates of increase occurred during the nineteenth century due to relatively high rates of immigration. While the number of colleges and universities fluctuated over much of that time period, their capacity grew to accommodate the students who wanted to attend. However, starting in the mid-1950s, with accelerating population growth and spiking demand for higher education, increases in the numbers and size of colleges had not expanded enough to keep up with the number of Americans seeking to enroll (Fischer and Hout 2006).

It is a truism of organization theory that as the number and scale of organizations increase, they become more differentiated (Blau 1970). This process occurs both at the level of the individual organization, which develops specialized components—for example, academic departments, administrative units, athletic programs, student services—and at the field level, where diverse types of organizational forms emerge. We described in chapter two the principal types of colleges identified by the Carnegie Foundation and depicted their development over recent decades in the Bay Area. Geiger (2011) provides a nuanced, historically based discussion of the differentiation process at the field level, with the emergence of colonial colleges prior to 1776; the development of the first state colleges after 1776; the growth of denominational colleges after 1820, focusing on the liberal arts and theological training; the origin of the land grant colleges in 1865, emphasizing utilitarian training in agriculture and engineering; the emergence and growth of community colleges after 1900; and the growth of for-profit colleges after 1975.

Misalignment

Another important development in higher education has been its increasing differentiation (separation) from the K–12 elementary-secondary educational system. When they first emerged in the beginning of the twentieth century, many of the developing junior colleges were outgrowths of local high schools (Geiger 2011: 56). These colleges were viewed as mechanisms for encouraging and enabling high school graduates to continue to pursue additional education. During the midcentury period of rapid growth, junior colleges were reconfigured as "community colleges," separated from K–12 districts and administered by their own governing boards operating at a regional and/or state level. Meanwhile, as high schools grew, they became differentiated; called "comprehensive high schools," they served complex and sometimes conflicting purposes and did not focus primarily on college preparation. College preparation classes were reserved for a minority of students in separate tracks of advanced placement (AP) or honors courses. Alternatively, this type of education took place in a distinct collection of elite "prep" schools, both independent and in the public "magnet" model. In recent decades in the United States, the chasm between secondary and postsecondary education has grown larger than in comparable programs in other industrialized nations (Clark 1985).

   Although organizationally and structurally distinct, colleges continue to be sequentially interdependent with the secondary schools from which most of their students come. Differentiated systems that are functionally interdependent require some mode of integration to be effective in carrying on shared work. The more complex the interdependence, the more elaborate the integration mechanisms (Lawrence and Lorsch 1967). Prior to the development of the comprehensive high school, US colleges and universities played an important role in attempting to better integrate the two levels. Starting in the 1890s the nascent regional accreditation boards and the National Education Association began setting standards for high school course-taking that would lead to college admission (VanOverbeke 2008). Soon thereafter, the regents of the

University of California began to accredit the state's high schools to make sure their curricula were adequate for university preparation, a regulatory structure that continues today in a modified form with California's "A-G" requirements, a collection of 15 yearlong high school classes required for admission to the state's public four-year colleges.

In the years after World War II, the notion of academic standards shared across the secondary and postsecondary sectors faded. Aptitude tests such as the SAT replaced subject matter standards for college admission, and secondary schools placed more emphasis on elective courses in nonacademic areas (Powell, Farrar, and Cohen, 1985). Currently, although secondary school and college faculty may share some of the same disciplinary interests, they rarely meet to discuss curricular alignment or to develop common standards. Many groups operating at local, regional, and state levels currently attempt to mediate between high schools and colleges, but they have competing agendas and often tend to work against curricular alignment. The number and influence of these mediating groups, such as the College Board (and its partner, the Educational Testing Service, which administers the SAT and AP exams), the American College Testing program, and the International Baccalaureate program, is an indicator of the "amount of disorder and confusion that has grown through the years in the relationship between the school and the university in America" (Stocking 1985: 261).

The starkest indicator of the severity of the misalignment between high schools and colleges is the number of college students requiring "remediation"— courses that enable high school graduates students to perform college-level work. For example, about 70 percent of students entering California's community colleges require some remediation work. Roughly 30 percent of all English and math courses offered by these two-year colleges are remedial level (Cohen and Brawer 2003). The majority of students enrolled in these courses are of traditional college age and entered college directly after graduation from high school. This implies that the high level of remediation is not just a result of having to refresh the skills of individuals who have been out of school for a time but a need to (re-)teach skills that were not learned in high school. A broader indicator of US students' inability to perform college work is evident in our low graduation rates. While almost 70 percent of high school graduates enroll in college within two years of graduating, only about 57 percent of students who enroll in a bachelor's degree program graduate within six years, and fewer than 25 percent of students who begin at a community college graduate with an associate degree within three years (Lewin 2010).

The most comprehensive recent effort to deal with the misalignment of high school and college work is the advent of the Common Core Standards movement in 2010. This program is described in chapter seven.

New Forms of Organizing

Silicon Valley did not independently invent the style of organizing for which it is now the poster child. These new forms emerged slowly over time in response to the growing worldwide pressures of globalization, the emergence of information technologies, and hypercompetition among providers that rendered older organizational forms inadequate. Dominant through the twentieth century, corporations had become larger, divided into divisional segments, hierarchical, and more stable as increasing numbers of functions were internalized to ensure adequate control by professional managers (see Scott and Davis 2007: chap. 13). But as new industries emerged, fueled by global competition and the need for customized, rapid response, new models of organizing began to appear.

Based on earlier craft models such as construction, publishing, and independent film making, these new models were marked by greater autonomy of work groups or teams, loose coupling of components, and reliance on shared work norms and trust engendered by personal networks (Powell 1990). Thoughtful observers also called attention to the emergence of regional districts made up of a dense, overlapping cluster of forms, such as those found in German industrial districts or in the collection of textile firms in the Emilia Romagna region of northern Italy (Piore and Sabel 1984). The competitive advantages of "flexible specialization" were becoming apparent. Leanness and agility were prized; horizontal integration among small-scale units promised quick responsiveness to changing demands; network forms held competitive advantage over hierarchical structures; alliances were preferred to acquisitions (Child 2005; Malone, Laubacher, and Morton 2003). These early outliers provided the formulae for the organizational models adopted by Silicon Valley firms, as described in chapter three.

Political and Economic Forces Shaping Higher Education

The social historian Christopher Loss argues that higher education institutions have played a central role in mediating between a relatively weak American federal state and its citizenry. In a number of important initiatives, partnerships were forged between federal and state governments and between public and private organizations and professional groups to advance a common agenda. Over the last century, American colleges have been pivotal institutional players in advancing public policy, including mobilizing the nation during the world wars, implementing New Deal programs during the Great Depression, and continuing into the postwar period in which diversity issues became dominant (Loss 2012).

For example, during World War II and the later Cold War, the federal government entered into vital partnerships with higher education, including both public and private institutions, to support research and training in areas of national interest like science, engineering, and the new interdisciplinary "area studies" fields. More broadly, the convergence of social movements pressing for reform and the actions of Democratic presidential administrations during the 1940s and 1960s expanded educational opportunities for all Americans. These initiatives largely had the goal of improving access to higher education for qualified students. From the GI Bill in 1944, providing financial assistance to demobilized servicemen to pursue a college education, to the Higher Education Act of 1964, which promoted statewide plans and provisions for expanding higher education throughout the nation, to the federal student aid programs that provide loans and grants to college students, the federal government has aided students seeking a college education (Cole 2010).

Our study period, beginning in 1970, highlights the time in which the federal government championed increasing access and equity in education, especially through individual loans and grants; while state governments increased their level of support for higher education early in this period, they have recently reduced funding.

Access and Equity

Americans have long emphasized the value of equality of opportunity and, as college increasingly is viewed as an important gateway and channel toward a better social and economic life, efforts have been made to ensure that a college education is available to all qualified students. Zemsky (2009: chap. 7) reminds us of several kinds of barriers to this goal. For years, many students were deterred from college by racial, gender, and religious *discrimination*. Substantial progress has been made in combating these restrictions, although inequities based on ethnicity and religion remain. As we have seen, students from various ethnic backgrounds are differently sorted among the varying types of colleges.

A second barrier was presented by a *scarcity* of colleges and universities, a problem addressed by a large expansion in the numbers and types of colleges beginning in the 1960s. American higher education experienced extraordinary growth during the twentieth century. The United States built the largest and most comprehensive public educational systems in the world with the expansion of both K–12 and higher education during this period. Access to college education expanded from less than 5 percent of the college-age population attending in 1900 to 2000, when more than 70 percent of youth attended college and over 30 percent earned a college degree (Fischer and Hout 2006). As Fischer and Hout (2006: 251) point out, "At the end of World War II, there was space enough in colleges and universities for only one-fifth of Americans age eighteen to twenty-two; by the early 1990s, there was space enough for about four-fifths of them. The expansion came about almost solely through the construction and subsidization of *public* higher education." Most of this expansion, particularly in the fast-growing states of California, Texas, and Florida, occurred through growth in the number of community colleges, which provided a combination of remedial education, vocational preparation, and liberal arts education to prepare students for transfer to four-year colleges (Marcus 2005; see also chapter two). In the process, as described above, community colleges became more detached from secondary education and took on new challenges, including vocational and adult education and community service. New and neglected populations apart from recent high school graduates were added, including housewives, immigrants, laid-off industrial workers, and older adults.

A final barrier, and one that has proved more difficult to overcome, is *financial*. For more than a half century, since the end of World War II, the average cost of a year in college has been increasing faster than the underlying rate of inflation. As displayed in table 4.6, the cost of annual tuition and required fees in current dollars has increased from roughly $1,000 in 1970–71 to over $30,000 in 2010–11 (National Center for Educational Statistics 2012). We recognize that listed tuition is often higher than the amount paid by students, but it nevertheless provides a useful indicator of the magnitude of increases over time. Costs for universities were about twice as high as for two-year programs, and private colleges and universities, both non- and for-profit, were nearly three times as high. Increasingly, these increased costs are borne by individual students and their families. Students have been forced to borrow larger sums,

Table 4.6 Average Undergraduate Tuition and Fees in Current Dollars Charged for Full-Time Students in Degree-Granting Institutions in the United States, by Level and Control, 1970–71 to 2010–11

	Year	4-Year universities	Other 4-Year institutions	2-Year institutions
Public	1970–71	1,362	1,135	951
	1980–81	2,712	2,421	2,027
	1990–91	5,585	5,004	3,467
	2000–01	9,948	8,715	5,137
	2010–11	17,722	14,979	8,085
Private	1970–71	3,163	2,599	2,103
	1980–81	6,569	5,249	4,303
	1990–91	16,503	12,220	9,302
	2000–01	29,115	21,220	15,825
	2010–11	46,519	30,071	23,871

Source: Digest of Educational Statistics, 2012, Table 349, National Center for Education Statistics.

and the increased burden of debt has become an issue of national concern. Students receiving an undergraduate degree in 2012 owed an average of over $29,000 (*Economist* 2014a).

Major Policy Changes

While education is the responsibility of states and localities, the federal government has played an increasing role in the sector since the mid-twentieth century. A chronological analysis of the major policies influencing postsecondary education in the United States and California, specifically, reveals three major trends in federal and state policy: (1) a constant focus on expanding access to postsecondary education opportunities to meet student and market demands; (2) attempts to address the issue of increasing educational costs, although state efforts have been erratic and often inadequate, while federal efforts have failed to keep pace; and (3) the increasing effort to hold educational institutions and systems accountable for meeting student and market needs.

Federal Structures and Policies

In the United States, the federal government from the outset has played a secondary role to that of the states, although over time its involvement and influence have substantially increased. Still, up to the present, its efforts in higher education are severely fragmented, being channeled through a wide variety of agencies—including the Departments of Education, Defense, Labor, Agriculture, Homeland Security, Transportation, Health and Human Services, Veterans Affairs, and the National Aeronautics and Space Administration—each with varying agendas and modes of intervention (Mumper et al. 2011).

In general, the period during and immediately after World War II witnessed a period of significant growth in the federal government's involvement in postsecondary education. As early as 1944, with the passage of the Servicemen's Readjustment Act, popularly known as the GI Bill, hundreds of thousands of veterans qualified for taxpayer-funded assistance to attend college. During the Cold War, federal funds provided grants and contracts to research universities to create the scientific and technical muscle to secure America's dominance over the Soviet Union in space and defense. The federal role increased and expanded with the emergence of the Great Society programs, and principally aimed to increase educational opportunities for all students and to redress racial and income inequalities. These goals were advanced by programs such as the Civil Rights Act of 1964, the Individuals with Disabilities Education Act of 1975, and the Higher Education Act of 1965. In 1979, legislation created the Department of Education (DOE) as a cabinet-level department and solidified the role of the federal government in setting standards for higher education as well as removing barriers, especially cost barriers, that kept low-income students from attending (Mumper et al. 2011).

During the Cold War, the Department of Defense along with the National Institutes of Health poured millions of dollars into research universities and comprehensive colleges, but such targeted support has been in decline since the 1990s. By the first decade of the twenty-first century, federal assistance to higher education was roughly evenly divided between the amount expended on direct student assistance, including student loans, veterans' benefits, and tax credits for students and families (43 percent), on the one hand, and funding for research (45 percent), on the other (Mumper et al. 2011).

Federal Student Aid. Motivated by the success of the GI Bill, the DOE extended direct financial support to a much larger range of students. The centerpiece of the program, the Pell Grant, which provided need-based grants to students, appeared in its present form in 1972. Earlier programs included both grants and loans, but over time federal financial policy gradually shifted to emphasize loans over grants (Thelin and Gasman 2010). Earlier approaches had "channeled aid to institutions, not individuals, and relied on aid officers to compile 'aid packages' using a combination of work-study, grants, and loans" (Loss 2012: 211). Framed by the conservative politics of the Nixon administration, which favored market mechanisms, federal assistance was provided directly to individuals, allowing them to choose among colleges with the loans supported by a financial framework combining public agencies with private banks.

The federal government continued to expand access to postsecondary education in the twenty-first century, primarily through the Post-9/11 GI Bill and the American Recovery and Reinvestment Act, which provided increased funding for student financial aid and research. The latter act also provided considerable relief to states, whose educational funding was threatened by the credit market crisis of 2009, by propping up state budgets to reduce the impact of the financial crisis (Mumper et al. 2011).

Although the number of Pell Grant recipients has expanded—from 1.9 million in 1977 to 5.5 million in 2007, the increase in the size of Pell Grants has failed to keep pace with increasing tuition costs. In the 1970s, Pell Grants covered more than 80 percent of average public university tuition but in 2012 covered less than a third (Kamenetz 2010: 61; see also Kamenetz 2008). As discussed above, these developments have led to increased debt obligations for a large number of students.

Practical Education and Workforce Development. In the midst of the Civil War, the federal government took its first massive step to influence the development of higher education with the passage of the Morrill Land-Grant Act of 1862, partnering with the states to create a set of public institutions devoted to linking higher education with the "applied arts" of agriculture, engineering, and technology (see chapter two). At about the same time, the US Department of Agriculture was created to advance a parallel agenda "to produce and disseminate up-to-date information and education on farming techniques, home economics and household management . . . to farmers and their families" (Loss 2012: 58). In 1914, these efforts were consolidated and expanded with the creation of the USDA Extension Service to partner with the land-grant colleges to conduct research and extend educational programs to enhance agricultural productivity, creating "one of the major educational agencies in the United States" (Loss 2012: 55).

During the 1960s, the federal government invested in workforce development programs, such as the Vocational Education Act of 1963, which provided matching federal funds to states to create vocational training programs in agriculture, home economics, and other trades. These efforts were expanded in the early 1970s with the passage of the Comprehensive Employment and Training Act (CETA), which allocated funds to state governments to provide on-the-job and in-class training to low-income individuals or to those who had been unemployed for a long period.

In the 1990s, the Perkins Vocational and Applied Technology Education Act provided funding to offer grants and create robust academic, occupational, and technical skills programs for secondary and postsecondary students enrolled in vocational and technical education. Later legislation included the School-to-Work Opportunities Act in 1994, which allocated funds to improve students' school-to-work transition by providing vocational and academic education, and the Workforce Investment Act of 1998, which provided funding for adult educational and vocational programs, as well as programs to increase the educational attainment and employment of students from disadvantaged backgrounds (Gordon 1999). A number of community colleges have begun to identify the most promising employment opportunities and have taken steps to provide curricula to address these needs (Grubb 2001).

Another approach to expanding the nation's educated workforce was the passage of the Immigration Act of 1990, which increased the supply of H-1B visas to allow US companies to employ for limited periods foreign workers who qualified for occupations requiring "highly specialized knowledge." Also, through the visa program, foreign workers who received a master's or higher degree from an educational institution within the United States were exempted from the program's quota.

Expanding the Field Boundaries. As Peterson (2007: 158) argues, in addition to shifting federal support primarily away from institutions to individuals (students), the 1972 Higher Education Act (HEA) "almost immediately created a new conception of industry," or in our terms a new conception of the organization field—a broadening from "higher" education to "postsecondary" education. In addition to the traditional degree-granting colleges, specialized vocational and technical training programs that were primarily for-profit enterprises became eligible to compete for students receiving federal funding. At the time, this change added almost 7,000 new institutions to the higher education field and more than doubled the number of students who were potentially eligible for federal student aid (Carnegie Commission 1973). By shifting financial aid from institutions to individuals and by expanding the types of institutions eligible for financial assistance, the 1972 HEA substantially enhanced competitive pressures within the field.

Because of this policy change, proprietary schools and colleges that had long operated on the margins of the field began to move into a more prominent position. The for-profit sector saw the opportunity for quick profits. Large, publicly traded companies, such as the Apollo Group, Corinthian, and DeVry, were quick to take advantage of the new funding stream and by the 1990s were offering liberal arts degrees as well as certificates. Over the past three decades, enrollments in for-profit colleges offering degree programs grew at about seven times the rate of the entire postsecondary sector (Hentschke, Lechuga, and Tierney 2010; Tierney and Hentschke 2007).

Because the for-profit sector is almost exclusively dependent on revenues from tuition, some companies engaged in aggressive marketing techniques and made inflated promises to student applicants about the successful placement of graduates. Beginning in the 1990s and continuing up to the present, these practices were met with heightened regulatory scrutiny from the DOE, congressional committees, and the US Department of Justice, which have investigated charges of fraud, recruitment violations, noncompliance with the requirements of Pell Grants, and false claims regarding future jobs. These accusations have not only been leveled at marginal "fly-by-night" operators but also involve a number of the mainstream for-profit companies, including ITT Educational Services, Corinthian Colleges, and the University of Phoenix (Breneman, Pusser, and Turner 2006; Tierney and Hentschke 2007) (see case example 5.A).

Although for some purposes the boundaries of the field of higher education have been broadened to include a wide range of specialized and for-profit postsecondary programs, the development of appropriate governance structures and mechanisms has lagged behind. As discussed in chapter two, most of the official data-gathering and research agencies continue to restrict their attention to degree-granting programs, meaning that there is only limited information about these other players (see appendix B). And most states have not effectively incorporated oversight and integration of these programs into their educational policy and planning activities.

More generally, the broadened boundaries represent a break in the sovereign authority long exercised by professional associations within higher education to control the standards governing the provision of postsecondary education in this country. These professional bodies have experienced reduced power in controlling such decisions as who is qualified to teach, what courses are to be offered, and what types of colleges and programs qualify for federal funding. In comparative terms, higher education appears to be in transition between being a field broadly subject to unified professional control, such as medicine in the United States, to one in which their jurisdiction is, at best, circum- scribed, so that broad segments of the field operate beyond their oversight. A possible comparison is to the field of mental health, in which roughly 20 percent of mental health services are provided by practitioners under the control of mental health specialists (e.g., psychiatrists, psychologists), whereas most are serviced by nonspecialized providers, including religious, criminal justice, and social welfare personnel (Koran 1981; Scott 1985).

Accountability. Federal oversight necessarily follows for any organization that receives federal funds. In part because of the large number of specific agencies supplying funds and the wide range of specific programs adminis- trated, the accounting burden falling on colleges participating in federal pro- grams is heavy (Scott and Meyer 1992: 147). Each agency has its own rules and reporting requirements. With the advent of civil rights programs, beginning with the Civil Rights Act of 1965, federal authorities—through legislation, requirements developed by regulatory agencies, and judicial decisions—have played a major role in the oversight of who colleges admit, how they award scholarships, and how they hire and promote faculty and staff. Indeed, Loss (2012) argues that since the 1960s colleges and universities have become the major instruments through which the federal government has advanced its diversity and equity agenda. In addition, the courts oversee colleges in such areas as the protection of free speech, separation of church and state, freedom of information, patenting of inventions, and protection of intellectual prop- erty (Olivas and Baez 2011).

More recently, the federal government has begun to scrutinize colleges in central and sensitive areas, asking questions about their productivity, academic performance, and contributions to the economy. Studies have shown that graduation rates have declined since the 1970s. Over 58 percent of high school students entering college in 1972 graduated in four years, while in the 1992 cohort, only 44 percent did so (Bound, Lovenheim, and Turner 2010). How- ever, numerous observers have noted the limitation of focusing on graduation rates. Such metrics do not take into account the great variety of educational programs that comprise higher education—the varying missions of broad- access colleges and the changeable trajectories of students. Students enroll in college for many reasons, including to obtain a short-term certificate or to ac- quire specific skills that do not lead to a degree. Students transferring between institutions are not tracked in these statistics. In addition, as higher percentages of high school graduates have sought college education, levels of preparedness for college courses have fallen. Current efforts to evaluate college performance fail to adjust outcome measures for these differences in entering student characteristics.

In addition to examining graduation rates, DOE leaders have begun to call for better testing of "learning outcomes" of college students. The most concerted effort to advance this agenda occurred in 2005 with the work of the Spellings Commission. Following President George W. Bush's successful efforts to introduce national testing into K–12 schools, his secretary of education, Margaret Spellings, organized the Commission on the Future of Higher Education whose recommendations included the development of improved performance measures. While no national system was put in place as a consequence, the commission's work is credited with documenting the deficiencies of current practices and stimulating increased attention to improvement (Zemsky 2009: chap. 1).

The popular media have also increasingly begun to collect and report data on college performance. In addition to graduation rates, reviewers such as *U.S. News and World Report* provide data on freshmen retention rates. Others are pressing for more attention to market-based measures, for example, assessing wages earned by graduates, using wages as a proxy for skills. All of these measures have their limitations, but there is little doubt that colleges are facing stronger accountability pressures in today's competitive climate (Arum and Roksa 2015).

State Structures and Policies

States have long held the major responsibility for educational policies in the United States, but they were slow to become involved in higher education. Most states assign responsibilities for the governance of colleges and universities to one or more boards, often composed of laypersons representing the public interests, but the specific administrative mechanisms employed vary greatly among the states (McGuinness 2011).

California played a leadership role early on among states in developing its innovative Master Plan, adopted in 1960 to bring a publicly funded college experience within reach of all qualified high school graduates. This plan, the brainchild of Clark Kerr, then president of the University of California, created a three-tiered framework for higher education, including the University of California (UC) system, the California State Universities (CSU), and the

California Community Colleges (CCC)—and assigning each its own distinctive mission and pool of students. The UCs, which today include 10 campuses, were designated California's primary research universities, while the CSUs, now numbering 23 campuses, were to provide undergraduate and graduate education through the master's level, including professional education. The objectives of the CCCs, which today number 112 campuses, were more diverse, being charged with providing vocational and academic instruction for any state resident wishing to pursue postsecondary education, as well as providing remedial instruction, workforce training, adult education, and course work able to transfer to four-year colleges. By providing strong guidelines for the mission of each tier, the plan avoided the high costs of isomorphism that would have come from all public institutions attempting to provide the same types of educational programs.

In spite of what was hailed as a superior design for the development of a comprehensive system of higher education, in operation the Master Plan created individualized policies for each tier rather than a framework for coordinating efforts to support statewide educational improvement (Finney, Riso, Orosz, and Boland 2014). In addition, the level of control exercised by the state varied. It granted the UC system a large measure of autonomy and protected its claim to state resources as the state's primary research university system, but it served to reduce access for California students to both the higher two tiers (Douglass 2010). In 1973, the state attempted to resurrect its role as a leader in higher education policy by creating the California Postsecondary Education Commission (CPEC). CPEC was charged with coordinating the state's higher education policy and practice, as well as improving integration with private sector developments. However, the commission failed to develop a comprehensive approach to deal with the rising number of students demanding postsecondary education or to foster better integration across types of colleges. Rather, the commission increasingly focused on the individual tiers and sectors and advanced only short-term solutions. Viewed as ineffective, CPEC was shut in 2011.

Once hailed as a leader among states, California now ranks near the bottom in such criteria as access to higher education and degree production (Douglass 2010). Recessions have brought financial stringency and severe restrictions to college access, principally in the CSU and CCC systems. Colleges have responded by increasing tuition and fees in all three tiers (Callan 2009), as described below.

State Funding: Colleges. The main support for public colleges has always been and continues to be the states, but state funding for colleges has dropped from covering over 50 percent of the expenses of these institutions to under 30 percent in 2012 (National Center for Educational Statistics 2012). In addition, the level of state funding is volatile. Because higher education is often one of the largest discretionary items in a state's budget, funding for it tends to rise and fall with the economy, or with policy changes that affect funding streams (Zumeta et al. 2012). For example, in California, the passage of Proposition 13 in 1978, the People's Initiative to Limit Property Taxation, capped annual property taxes at 1 percent of a property's value through the duration of an individual's ownership. As a result, Proposition 13 has placed economic constraints on the Master Plan's original commitment to provide a postsecondary education to all qualified Californians (Callan 2009). Moreover, the passage of this proposition required California localities to rely on other sources of educational revenue that were less reliable than the property tax base, such as sales and income taxes.

Throughout the 1980s and 1990s, the state lacked a comprehensive, long-term financial plan for higher education, instead focusing on the individual tiers of colleges and short-term solutions. For example, for a time, compacts between California's governor and the UC and CUS tiers provided specified levels of support in return for specified levels of performance—primarily involving the admission levels for in-state students and controls on tuition and fee increases (Finney et al. 2014). These compacts broke down during the economic downturn in the early 2000s (Douglass 2011).

The passage of Proposition 98 in 1988 guaranteed a minimum funding floor for CCCs and K–12 public education, providing some security for the community colleges. However, CCCs, unlike four-year schools, have a very low tuition and do not have other revenue sources, such as endowments or research grants, to fall back on. During lean budget years, they have a limited ability to meet their multiple missions and accommodate the increasing number of students seeking admission.

The trends during our period of study show a story of decline. Data compiled from the California Department of Finance reveal that the overall share of general fund expenditures allocated to higher education has decreased from 18 percent in 1976–77 to under 12 percent in 2012–13 (Randolph and Johnson 2014: 14). A longitudinal look at state support for the three tiers of public colleges in California for the same period shows some volatility but substantial overall decline in funding for colleges (fig. 4.3). General fund appropriations for FTE students, adjusted for inflation, declined particularly for the UC system, less so for CSU, and, although funding is modest, have remained fairly steady for CCCs.

Confronted with diminishing state support, public colleges have been obliged to respond through cost-cutting measures—including increasing class size, reducing class offerings, reducing services, decreasing faculty hires—and raising student fees and tuition. Tuition and fees began to rise in the early 1990s and have accelerated their rates of increase since 2000. At the same time, both UC and CSU have increased the proportion of their students who are either out-of-state or international, since they are able to charge these students substantially more than in-state students. Data compiled by the Office of the President, University of California, show that the gap between the number of California's high school graduates applying to the UC and CSU systems and those being admitted has doubled since 1996. An increasing number of students qualified for a selective college can find no room in any of the state's four-year public institutions (Campaign for College Opportunity 2015; see also fig. 2.4). A telling indicator of this failure is the increase in the number of

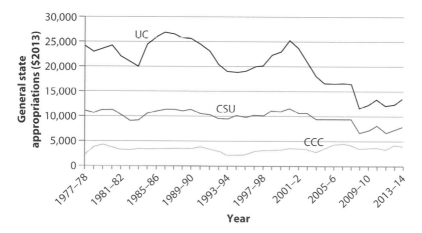

Figure 4.3. General state appropriations per full-time equivalent students by California's three college tiers, 1977–2014 (in 2013 dollars adjusted for inflation).
Source: California Postsecondary Education Commission "Fiscal Profiles 2010" Legislative Analyst's Office. This figure is used with permission from Randolph and Johnson (2014).
Note: The definition of an FTE student has changed over the years.

"bottlenecks" and "impacted" colleges or programs—a problem that also affects access to CCCs (see case example 4.B).

CASE EXAMPLE 4.B

BOTTLENECK AND IMPACTION IN THE CALIFORNIA STATE UNIVERSITIES

Years of budget shortfalls in the state's postsecondary education system have resulted in several problems at each of its three segments: the University of California (UC), the California State Universities (CSU), and the community colleges (CC). Chief among the problems in the CSU system are the related issues of bottleneck and impaction. A *bottleneck* is defined as anything that limits a student's ability to make progress toward a degree and graduate in a timely manner (e.g., being unable to enroll in a required course because it is overenrolled). The consequences of bottlenecks are varied and serious (e.g., delayed time to degree) and while bottlenecks exist across the CSU campuses, institutions in hi-tech regions (e.g., San Jose, Long Beach) experience more challenges than do those in less industrialized regions (e.g., Fresno, Dominguez Hills).

A related problem is the issue of *impaction*. An undergraduate program or major can be designated as "impacted" if it receives more applications from qualified applicants than it can accommodate, given limited resources and capacity (California State University 2002). While there are no assurances, applicants may still be admitted to the campus of their choice in an alternate, non-impacted major, or they may be admitted eventually to their oversubscribed major if they meet supplementary admission criteria (e.g., a higher GPA, more transferable course work from a local community college). Still, admission to a CSU campus does not guarantee admission to a particular major, especially if that major is impacted. From 2009 to 2014, CSU campuses declined roughly 140,000 eligible California students because of limited spots and budget cuts (Campaign for College Opportunity 2015). Some CSU campuses are entirely impacted because all their programs are enrolled to capacity (e.g., San Diego, Fullerton, San Luis Obispo). Institutions that have campus impaction must restrict enrollment by increasing selectivity in their admissions process. In fact, since 2004, 6 of the 23 CSUs have raised their admission standards by 135 percent, requiring applicants to meet a higher GPA and/or SAT score (Campaign for College Opportunity 2015).

Four Categories of Bottleneck

The CSU system identifies four types of problems that result in bottlenecks: (1) student readiness and curriculum, (2) place-bound, (3) facilities, and (4) advising and scheduling. The CSU Office of the Chancellor (http://courseredesign.csuprojects.org/wp/mission) is committed to increasing access, reducing time to degree, and improving success rates; it continues to employ many strategies to address persistent types of enrollment bottlenecks. Many of these interventions integrate technology (e.g., proven course redesign).

First, academic preparation can often be a formidable barrier to students' success in college. Many students in CSU are unprepared for college-level work—despite meeting CSU's admission requirements—which results in a large number withdrawing from or failing courses, and having to retake courses in order to graduate. Being unprepared for courses results in bottlenecks that are created both by the combined enrollment demands of new students and students needing to repeat courses. Recently, the CSU system identified courses that have high enrollment but low success. Thirty-six percent of these courses were in STEM-related disciplines such as chemistry, biology, calculus, and statistics. According to the state's Legislative Analyst Office, about 58 percent of regularly admitted freshmen in CSU in Fall 2009 were deemed unprepared for college-level writing or math (or both). As a result, many had to take some form of remedial course work, which, in turn, delayed their completion of a postsecondary degree.

To address remediation, CSU initiated the Early Assessment Program (EAP) in the early 2000s, a system to identify students who are not yet ready for college-level courses (in English and math) and link them with supports to improve their readiness (Howell, Kurlaender, and Grodsky 2010). The EAP has three components: an eleventh grade test to assess their level of preparedness for college-level work; professional development training for high school teachers to improve students' postsecondary readiness; and supplemental instructional supports in students' senior year to get them ready for college (e.g., access to online math programs).

Building on the success of EAP, the CSU system also initiated Early Start in 2012. This program requires freshmen who do not pass CSU's placement exams to begin remediation during the summer before taking their regular course work. Additionally, some campuses have pursued a variety of other strategies, including offering Summer Bridge programs to low-income, first-generation students, which provide remedial classes as well as college supports, counseling, and orientation.

In an effort to ease curriculum-related bottlenecks and improve students' success, the chancellor's office has implemented "Summer E-Academies," which aid CSU faculty to redesign their courses using technology. Faculty at San José State University, for example, have developed a "flipped classroom" for a popular engineering course (Engineering 098); other courses may also be redesigned so that they are fully delivered online to accommodate demands. Flipped classrooms, which have been in existence for over a decade, typically involve courses that move the traditional lecture or content dissemination to online platforms (e.g., video lectures), accessible to students outside of class time. The face-to-face class time is then used for application of concepts.

A second type of bottleneck is associated with students' local campuses. Frequently, students in California are place-bound and attend universities convenient to where they live. Because of this, students often must wait for their campus to schedule particular required courses. These place-bound bottlenecks are especially significant for students at smaller CSU campuses where a broad array of course requirements competes for limited resources. To ameliorate the problem, CSU has instituted the Intra-System Concurrent Enrollment (ICE) program, which allows students enrolled in any CSU campus to take courses at other CSU campuses, on a space-available basis—unless those campuses or programs are impacted. Increasing the number of online courses that articulate between institutions is perceived by many to be a promising approach to tackling enrollment and capacity demands.

The unavailability of campus facilities can also result in bottleneck. For example, enrollment demands have outpaced the physical capacity of many campuses to offer introductory STEM courses that have laboratory requirements (e.g., chemistry), thereby hindering the number of students who can take lab sections in safe and properly equipped facilities. To address this kind of bottleneck, CSU has initiated the Virtual Labs program—a systemwide academic technology initiative that creates computer-based activities in which students interact with an experimental apparatus or other activity via a computer interface. Virtual Labs can be integrated with in-class lectures through a hybrid-flipped lab model where students alternate conducting laboratory work in person and online; this, in turn, can help address the lack of laboratory space and allow more students to be served.

The last type of bottleneck is associated with information. Students are frequently unaware of the wider range of course and program options available to them to complete their general education and major requirements. These advising and scheduling bottlenecks result when students do not receive timely advice about their academic pathways and course schedules. To prevent such issues, the chancellor's office has urged campuses to leverage technology to provide students with clearer roadmaps to graduation. Proposed solutions have included both student- and institution-facing interventions such as enhanced degree audit systems (which enable students and their advisors to assess academic progress), web-based academic planning tools, early warning systems (which afford better case management of students at risk of not graduating), as well as predictive analytics that allow institutions to uncover insights and patterns about student achievement and track the efficacy of efforts designed to alleviate bottlenecks.

Although the CSU system has implemented a variety of strategies to address access issues, many observers contend that considerable updates to the Master Plan will be necessary to comprehensively address shortcomings in the state's overall postsecondary system (e.g., Douglas 2010; Finney et al. 2014).

State-Funded Student Aid. Another way the state government exercises authority over California higher education institutions is through student aid. The Cal Grant Program, administered by the California Student Aid Commission (CSAC), dates to 1955, and is only open to California students attending in-state institutions, both public and private. In 2000, Governor Gray Davis signed into law a bill that reorganized the program into five different grants. Three are entitlement programs for recent high school graduates, while two are competitive programs with a fixed number of awardees, meant to be nontraditional students. CSAC does not prominently advertise eligibility rules, instead encouraging all students to complete a unified, simple application process.

In 2012, Governor Jerry Brown signed new legislation that used the Cal Grant Program to exert increased control over higher education institutions through the guise of a budget cut. That year's state budget reduced spending on Cal Grants by $134 million as part of an effort to address a $16 billion overall shortfall. Accompanying the cut were new rules limiting eligibility to institutions with a six-year graduation rate of at least 30 percent and a maximum three-year cohort default rate on student loans of 15.5 percent. The new rules specifically exempted community colleges, causing them to apply almost exclusively to for-profit institutions. An initial report by the state Legislative Analyst's Office found that 154 institutions had been ruled ineligible for Cal Grants within two years of the law's passage; these included 80 percent of all for-profit colleges and universities in the state, including giants such as University of Phoenix (Legislative Analyst's Office 2013).

In 2015, researchers from the Public Policy Institute of California (PPIC) testified at a CSAC meeting that the structure of the Cal Grant Program did not encourage desired goals such as matriculation and completion at four-year institutions. Given the state's poor performance on these metrics, they suggested changes to Cal Grants that would incentivize students to take full course loads (15 units or more). Part of the plan would include increasing the value of awards to include living expenses so that students could attend four-year institutions and reduce the number of hours that they work. PPIC also argued that California should replicate the work of other states in creating a longitudinal statewide database to track Cal Grant recipients from high school through college and into the workforce in order to assess the efficacy of the program's various aspects (Johnson, Mejia, and Cook 2015).

The Cal Grant C award presents the opposite policy challenge. It is not directed at baccalaureate programs at all, instead offering up to $3,000 in tuition and fees for students pursuing "technical or career education" (CSAC 2016). However, eligibility rules for the award mean that its flexibility is diminished. The funds are only available for programs that are at least four months long, meaning that some short-term "reskilling" programs are ineligible. Furthermore, the award can only be used at an annually approved list of accredited institutions, meaning that recipients cannot use it to attend highly specialized training programs like Dev Bootcamp (described in case example 2.A).

State-Funded Student Aid for Workforce Training. As record numbers of high school students are applying to state colleges and universities, more are also receiving hands-on training in high-demand technical careers even before they earn their diplomas. The students, many beginning in the ninth grade, are in career pathways learning job skills alongside professionals in fields including aviation, healthcare, civil engineering, fashion design, tourism, and new media.

These pathways, which integrate academics with real-world work experience, are fueled by California's unprecedented investment in career technical education. The state is pouring more than $1.5 billion over five years into programs aimed at establishing and strengthening partnerships between K–12 schools, community colleges, and businesses to better prepare students for college and careers.

During 2014–16, the nearly $500 million Career Pathways Trust has awarded 79 school districts, county offices of education, community colleges, and charter schools grants of $583,000 to $15 million each. The state last year also approved an additional $900 million in Career Technical Education Incentive Grants, which is to be awarded over a three-year period to dozens of districts and other agencies to accelerate the development of new programs. About a half dozen smaller grant programs are also helping to drive California's unprecedented growth in career technical education. However, these initiatives are one-time state funds, and it is unclear whether they can be locally sustained. The California K–12 education code contains only a few general phrases about state CTE policy, and attempts to provide a more integrated K–12 and postsecondary focus are just beginning.

College Finances: Overview

Before leaving the discussion of financial support for colleges, we broaden the scope to include the full range of colleges, including nonprofits and for-profits. As expected, revenue sources vary considerably across types of colleges (table 4.7). In data for 2006 compiled by Weisbrod, Ballou, and Asch (2008) for the United States, four-year public colleges received roughly equal proportions of their revenue from state and federal appropriations and grants, whereas two-year public colleges depend primarily on state revenues. Tuition and fees are nearly twice as high in nonprofit institutions as in public colleges and more than three times higher in for-profits than in public colleges. Nonprofits receive a significant proportion of their revenues from private and corporate donations, which represents a federal and state subsidy, since donors benefit from the tax-deductibility of gifts to nonprofit organizations.

In response to federal and state cuts, colleges have shifted the financial burden to students in the form of increased fees and higher tuition. Figure 4.4 shows that the tuition revenue per student has doubled among public, nonprofit colleges and more than quadrupled in for-profit private institutions in

Table 4.7 Sources of Revenue of Colleges, by Ownership, 2006

	Total revenue (in millions)	Tuition & fees	Federal appropriations	State & local appropriations	Federal grants & contracts	State & local grants & contracts	Private gifts, grants & contracts	Endowment income	Sales, services of educ. activities	Sales auxiliary enterprises	Other sources
Public											
4-Year	$315.2	17.1%	18.5%	26.8%	13.0%	6.8%	2.7%	1.3%	—	9.1%	4.7%
2-Year	46.8	14.7	5.6	54.9	10.7	6.4	1.1	0.4	—	4.3	1.9
Nonprofit											
4-Year	93.1	31.4	8.2	5.8	13.0	4.8	12.7	—	8.7	9.9	5.6
2-Year	11	20.0	7.3	33.6	5.5	3.6	4.5	—	3.6	8.2	13.6
For-profit											
4-Year	29.8	68.8	16.1	2.7	—	—	0.3	—	4.4	5.0	2.7
2-Year	9	55.9	22.3	4.5	—	—	0.6	—	3.4	7.8	5.6

Source: Weisbrod, Ballous and Asch, 2008, Table 2.1, pg. 30; Authors' calculations from data in U.S. Department of Education National Center for Education Statistics 2007b.
Note: —=no data available.

Definitions of categories:

- Tuition and fees: all tuition and fees minus discounts.
- Federal and state and local government appropriations: revenue provided by legislative act, usually for operating expenses, not specific projects; includes general appropriations from a state.
- Federal and state and local grants and contracts: revenue from governmental bodies for specific programs and projects, including research and training.
- Private gifts, grants, and contracts: "Revenues from private donors for which no legal consideration is involved and from private contracts for specific goods and services provided to the funder as stipulation for receipt of the funds. Includes only those gifts, grants, and contracts that are directly related to instruction, research, public service, or other institutional purposes."
- Endowment income: includes unrestricted and restricted endowment income; does not include "gains spent for current operations, which are treated as transfers."
- Sales, services, of educational activities: revenue from selling goods or services "incidental to the conduct of instruction, research or public service." Examples: dairy products, film rentals, testing services, university presses.
- Sales auxiliary enterprises: revenues from "self-supporting activities" that provide a service to students, faculty or staff. "Examples are residence halls, food services, student health services, intercollegiate athletics, college unions, college stores, and movie theaters."
- Other sources: "examples are interest income and gains (net of losses) from investments of unrestricted current funds, miscellaneous rentals and sales, expired term endowments, and terminated annuity or life income agreements, if not material. Also includes revenues resulting from the sales and services of internal service departments to persons or agencies external to the institution (e.g., the sale of computer time)."

California over the period 1980–2010. For instance, the average tuition at a UC and CSU campus more than doubled, from $4,000 to $9,000 (Jackson 2014). In fact, public institutions have seen the largest increases in posted tuition as the funding model has shifted from primarily state-subsidized higher education to self-financed higher education, most of it in the form of student loans. We have noted that such financial shifts have led to a growing burden of student debt for the average college student (Kamenetz 2008).

Changing Institutional Logics: The Rise of For-Profit Colleges

After a period of substantial public investment in higher education from World War II through the Great Society programs of the Lyndon B. Johnson era, federal policies since the 1970s have generally moved to favor private over public support and market solutions over government-based governance mechanisms. Public support for state colleges has been declining, as we have seen, and students and their parents are bearing a larger and larger proportion of the costs of education. Media rankings play an ever larger role in oversight and increasingly influence the enrollment decisions of students and the educational decisions of college administrators.

For-profit colleges have also been on the rise. During the past three decades they have grown at a faster rate than any other type of educational entity. Offering a wide array of programs in areas such as business, information technology, and health care, their enrollments more than tripled between 2000 and 2010. As of 2010, proprietary colleges accounted for about 33 percent of all associate degrees in business, management, and marketing; 50 percent in computer science; and 23 percent in health services (Deming, Goldin, and Katz 2010). Their attraction stems from the ability to offer more responsive, flexible, and customized programs to students who are employed and attend classes in the evenings or on weekends, and to offer shorter-term programs with clear pathways toward particular occupational goals. They have also been in the forefront of employing computer-based instruction (Tierney and Hentschke 2007). These substantial advantages are, however, offset by performance deficiencies.

As described earlier, in 1972, many of those granting degrees became eligible to admit students with federal grants and loans. Because their prices are substantially higher than public colleges and their admission standards lower, these schools have experienced higher dropout and student loan default rates

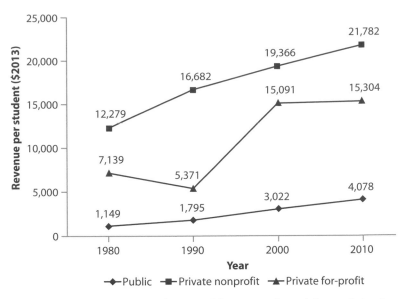

Figure 4.4. Tuition revenue per student in public, nonprofit, and for-profit institutions, 1980–2010.
Source: Higher Education General Information Survey (1970–1985) and Integrated Postsecondary Education Data System (1986–2012).

than comparable public programs. A 2012 congressional investigation found that, as a sector, for-profits had a 64 percent dropout rate and spent 22 percent of their revenues on marketing, advertising, recruiting, and admissions compared to 22 percent on teaching (*Economist* 2015a). Since 2010, enrollment at these institutions has been falling, likely due in part to the negative publicity that came from these statistics (see case example 5.C).

Technology-Driven Changes
Early Distance Learning Technologies

American higher education has a long history of experimenting with technology as a means to reach the broadest possible set of learners. We have described above the extension programs associated with land-grant colleges in the nineteenth century. By the 1920s and 1930s many of these programs, which sought to bring technical know-how and college-level courses to nontraditional students, readily adopted the radio as an instructional medium

(Craig 2000). Television was another natural instrument to exploit in teaching students who were not physically present on campus. It was widely touted as a way to reach a maximum number of students with maximum efficiency (Schramm 1962).

One of the most imaginative and successful programs to exploit the opportunities presented by the distance-spanning technologies of radio and television came from overseas: the Open University (OU), developed in the United Kingdom in the early 1970s. Guided by the vision of Harold Wilson, soon to become prime minister, the hope was to create a "university of the air . . . designed to provide an opportunity for those who, for one reason or another, have not been able to take advantage of higher education" (Wilson 1963). The OU began to offer courses in 1971 conducted through lectures broadcast on television and radio and materials sent by mail, including texts and "home experiment kits." Adults over 21, and later, 18 year olds, were freely allowed to enroll in undergraduate programs; course fees were modest. Enrollments grew steadily up until 1985 when they spiked with the opening of the university's school of business.

By 1990, the university had awarded 108,000 degrees, most at the bachelor of arts (BA) level, and had enrolled hundreds of thousands of others who had not completed a degree (Weinbren 2014). Apart from surface mail, the primary mode of instruction for the OU's first three decades was television, often produced in collaboration with the BBC. By the mid-1990s, as television's primacy was fading, some students began receiving interactive CD-ROM course work. Since 1999, an increasing number of courses have become internet-based. Online technology enabled the university to extend its reach even further. As of 2015, the OU has served nearly 2 million students and awarded over 380,000 BA degrees (Open University 2014).

Some of that reach has been on this side of the Atlantic. In 1999, the OU inaugurated a branch in the United States, with physical headquarters in Colorado. Catering primarily to community college graduates, it initially offered only the final two years of bachelor's degree work. However, this effort failed to attract sufficient students and folded after three years (Kirp 2004). More recently, the OU has sought and received accreditation from the Middle States Commission (one of the regional accreditation associations), and continues its efforts to attract nontraditional students in the United States.

In the Bay Area, Stanford University's School of Engineering was at the forefront of experimenting with distance learning. The effort began by first moving students rather than classes: in 1954 the school opened its on-campus Honors Cooperative Program (HCP), offering math, science, and engineering courses to full-time employees of selected local firms, as described in case example 3.A. These efforts were coupled in 1969 with connections bringing instruction into off-site workplaces in nearby companies via closed-circuit television feeds. In the early 2000s, these courses were connected to the internet. Some of the courses had restricted admission and led to advanced degrees, but most are linked to professional certificates.

Digital Education Arrives: A Redefinition of the Field of Higher Education?

Futurists have been predicting the advent and rapid acceleration of digital instruction for over three decades. This discourse entered a hyperphase in the past few years as a number of high-profile universities threw their hats into the digital ring by hosting massive open online courses, commonly called MOOCs (see case example 4.C). With the advent of OpenCourse-Ware, Coursera, Udacity, and other platforms, commentators and critics have been forecasting the massive reshaping, if not the end, of higher education as we know it, because of the "online tsunami" of e-learning (Carey 2015). Techno-utopians have touted the benefits of digital instruction just as earlier advocates insisted that our educational needs would at last be met by the advent of the miracle of radio, and then of television. The vision has been particularly seductive because, for too many qualified students, traditional educational systems have been unable to cope with increasing demand and rising costs. Early course offerings met with enormous success. In fall 2011, 160,000 students in 190 countries enrolled in an artificial intelligence course developed by a Stanford professor and offered for free by Udacity (Lewin 2012). However, to this point, results stemming from the introduction of MOOCs have been disappointing. While large numbers of students have enrolled in these courses, only a small proportion has completed them. And rather than serving the educationally marginal or disadvantaged, most of the students benefiting from these programs are those already well-educated, male, and currently employed (Alcorn, Christensen, and Emanuel 2014).

CASE EXAMPLE 4.C

MASSIVE OPEN ONLINE COURSES (COURSERA, UDACITY, AND EDX)

In 2001, MIT initiated OpenCourseWare (OCW), which catalyzed the field by offering abbreviated versions of traditional courses. Many viewed OCW as a promising model that could be widely replicated. In 2002, UNESCO devised the term *open education resources* to "include open applications, tools and architecture, as well as legal enablers like Creative Commons that allow the spirit and practice of open education to be exercised" (Fizz 2012). Selective universities, including Stanford and Harvard, contributed to the momentum of online learning, expanding it today to include the growing number and various forms of massive open online courses (MOOCs) (Johnson 2014).

In higher education, the current top three providers of MOOCs are edX, Coursera, and Udacity. Coursera and Udacity were created in Silicon Valley, and while these providers share certain characteristics, they also have distinct features and specialties (table CE4.C).

Started by two Stanford University computer science professors in April 2012, Coursera is considered to be the biggest provider of MOOCs. Coursera is a for-profit organization that generates revenue through verified certification fees and tuition fees. Coursera has focused their efforts on getting colleges and universities to approve their online courses for college credits. Some institutions adopting these programs include major US research universities such as Stanford, Duke, and the University of Michigan, and over 100 postsecondary institutions around the world. As of May 2015, Coursera has offered over 1,000 courses and has approximately 13 million users in 190 countries.

Founded in 2012 by Sebastian Thrun, a computer science and electrical engineering professor at Stanford University, Udacity is a for-profit business, with approximately 1.6 million users and 26 free courses. Of late, however, Udacity has pivoted its strategy from focusing on university-level courses to emphasizing vocational courses for working professionals.

A partnership of the Massachusetts Institute of Technology and Harvard University launched edX in May 2012. Supplying a wide range of courses, edX differs from other MOOC providers in that it is a nonprofit organization that uses open-source software. Currently, edX has approximately four million students worldwide taking more than 500 courses. It is also involved in research,

Table CE4.C   Pros and Cons of Online Course Providers

Pros and Cons	Coursera	Udacity	edX
Pros	• Free unverified certificates of completion. • Biggest catalog. • The best of the discussion forums but still limited. • The greatest variety of partners. • Transcripts in different languages. • iOS, Android and Kindle Fire apps. • Free "honor system" certificates; fee-based verified certificates; advanced certificates for specialized work.	• Start anytime. No waiting for the course you're interested in. • Move at your own pace. Caters to self-directed students. • Many programming and computer science classes. • More focus on current workplace skills. • iOS and Android apps.	• Free unverified certificates. • Big catalog from prestigious university partners. • Strong in sciences and medicine. • Some foreign language classes. • Free "honor system" certificates; fee-based verified certificates; advanced certificates for specialized work.
Cons	• Too structured for some learners. • More variability in quality. • Courses not available on demand.	• Fee-based certificates. • Smaller community. Most students work alone, unless they pay for a premium version. • Few offerings in foreign languages.	• Frustrating discussion forums. • More variability in quality. • No apps available.

Source: skilledup.com.

collecting student data to look for ways to improve course completion and student retention. Similar to other nonprofit organizations, one of edX's biggest challenges is determining how to sustain itself. The organization understands that it needs to raise enough funds to cover costs, including compensation for its partners.

The continuous growth of institutions providing MOOCs reflect the strong demand from users. Those who support MOOCs believe that higher education needs a tool that is on a scale wide enough to reach all students anytime and anywhere. They see MOOCs as helping to break the traditional mold of higher education by increasing access at low to no cost. One of the advantages that MOOCs possess is that they are designed to be self-paced; students can access the course whenever it is most convenient, allowing them to start and finish at their own pace. This is especially beneficial for students who have jobs, have to take care of children or other family members, or who are unable to come to a physical campus to take classes. Many of these students have historically lacked access to higher education due to their socioeconomic status.

Universities partnering with MOOC providers have seen positive impact. For example, "San Jose State University last fall [2012] used an edX MOOC, 'Circuits & Electronics,' as the basis for a "blended" online course offered to 85 tuition-paying students on its campus. The early results were encouraging. The semester before the edX pilot, 60 percent of students passed the San José State course; 91 percent passed the edX-infused version" (Kolowich 2013).

In the ecology of higher education, MOOCs play a specific role that provides inexpensive and efficient education to any student with access to the internet. Traditional brick-and-mortar colleges and universities will need to adapt to the ever-changing ecology of higher education. Thille, Mitchell, and Stevens (2015) argue that while these new types of courses are not a panacea, they are still important components of today's higher education field: "MOOCs have not fixed higher education, but they are poignant reminders of the urgent problems of college cost and access, potential forerunners of truly effective educational technology, and valuable tools for advancing the science of learning. That's progress" (Thille, Mitchell, and Stevens 2015).

Bay Area colleges from Stanford to San José State University to Foothill Community College have all taken steps to include online instruction in their portfolio of course offerings, and significant numbers of students, both remedial and advanced, have been able to take advantage of these resources. However, up to this point, neither these nor related efforts by other colleges have succeeded in changing the basic vocabulary of a "college education." Most colleges continue to carry on business as usual, the important exception being the for-profit colleges. They have been quick to embrace online courses. The new technology perfectly suits their production model, which relies on delivering courses to large numbers of students in efficient, cost-saving ways. They confront few of the barriers that hamper public and nonprofit colleges. They are not staffed by large numbers of tenured faculty that have a strong say in curricular and teaching decisions. And they confront fewer rules and regulations regarding course and curricular changes than those confronted by public colleges. Moreover, the raison d'être of for-profit colleges is to be responsive to consumer needs and preferences—to rapidly adjust their offerings as demands change.

Those who anticipate a revolution precipitated by online education are not simply discussing an improved method of providing education but are proposing a paradigm shift in the way education is understood. Rather than adopting the premise underlying our field approach that any analysis should begin with the existing system of educational providers—a supply side view—they instead propose that we view the field as a market in which the consumers (students) are the central actors of interest. Such a perspective shifts attention from providers to consumers, from supply-side to demand-side concerns and forces. Their approach carries echoes of concerns voiced several decades ago by an outspoken critic of educational organization. Ivan Illich (1972) argued eloquently if somewhat caustically that we should never equate education with schools; indeed, it would be a service to all, he said, if society could be "deschooled."

When they appeared half a century ago, Illich's views seemed, at best, a utopian mirage, but now in the second decade of the twenty-first century, they begin to emerge as a real possibility to many. The earlier alternatives to schooling—including experience, travel, and reading—have been vastly expanded by the internet as a source of information and potential learning. A variety of search engines are available to guide those in search of data, information, even knowledge on a vast variety of topics. And with the availability of a wide-range of college-level courses offered freely to all those with access to the internet, opportunities for learning and study are no longer restricted to those attending brick-and-mortar schools. Hence, enthusiasts advocate that students can now "design [their] own university" (Kamenetz 2010). Kamenetz describes some approaches under development to support liberated, independent learners, whom she pictures at the center of unlimited intellectual resources, supported by "community- and practice-based learning" (109). They have at their disposal resources designed by such providers as the Open University, MIT's OpenCourseWare, and edX, to enable them to construct their own curriculum. Online resources can be combined with self-organized groups of learners, both co-located and virtual. The model envisions bypassing all of the existing organizational paraphernalia of higher education: the gatekeepers, the 50-minute lectures at a set time and place, the prerequisites and requirements, the end-of-term examinations, and, most importantly, the high costs.

As argued elsewhere (Scott 2015), this approach, while intriguing and appealing in many ways, appears to posit a heroic, and no doubt unrealistic, portrait of the agency, maturity, and judgment of individual learners to chart their own course and mobilize the relevant intellectual supports. Indeed, the skills and discipline required to manage these tasks are precisely those associated with a highly educated person (see also Rosenbaum, Deil-Amen, and Person 2006). Note that, to date, most of the students benefitting from MOOCs and similar offerings already have college degrees.

Simon (1977 [1945]) long ago taught us about the cognitive limitations of individuals, pointing out that the primary purpose of creating organizations was to simplify complex work by subdividing it and to provide support for individual decision making by creating formal channels to supply appropriate information to participants. In short, he argued that organization is required to overcome the limits of individual rationality. Students benefit from being embedded in organizational systems of learning—"colleges"—for cognitive as well as social reasons.

And, indeed, it appears that new types of organization are beginning to emerge to enable individual learners (hopefully not to exploit naïve consumers) and to take advantage of new business models and income streams. Existing universities and colleges that have developed online courses are, as of now, unsure how to secure revenues from their offerings. And all of these developments raise a host of questions regarding oversight and governance: how to guarantee quality and certify competence. We have previously discussed some of the ongoing efforts to solve the validation/certification problem for new types of training programs and credentials (see chapter three and case example 3.C).

More generally, while new technologies and delivery systems are under development and assessment, they are all confronted by a highly institutionalized system of educational provision and certification. In short, the most salient environment for these new players is the existing field of higher education.

These new initiatives do not enter an empty space but an established, complex organizational field of educational programs and supporting entities. Given the familiarity and widespread acceptance of these structures and processes, it would be surprising if newer types of providers and approaches did not learn from, adapt to, and develop connections with existing entities. Institutional change involves a confrontation between the old and the new forms and ways. In most cases, through bricolage—complex processes of learning and compromise, of combination and recombination—new hybrid institutional structures emerge (Scott 2014: chap. 6; Stark 1996).

Concluding Comment

All organization fields are subsumed under wider social systems—state, national, and, increasingly, international. The field of higher education, more so than most other types of fields, has been expected to be responsive to changes occurring at these levels in demographic size and composition, political beliefs and policies, economic forces, and technological changes.

Demographic changes have been particularly challenging as the population in the San Francisco Bay Area has doubled between 1970 and 2010. Much of this growth has been fueled by immigration from Mexico and the Asia-Pacific regions, with the result that whites now make up just over 40 percent of the population. During the period of our study, the federal government was starting to play a stronger role in higher education, although the states continue to be the primary actors in educational matters, both K–12 and beyond. The 1960s and 1970s saw heightened attention to the issues of ethnic and income inequality, and these were translated for educational systems into pressures to provide better access to higher education for all qualified students. The pressures were national, but the bulk of the responsibility for expanding college systems fell on the states. During this same period, the major economic trends were toward the privatization of higher education. Increasingly, students and their families were expected to pay for a higher proportion of the costs of their education, colleges became more responsive to market pressures, and many for-profit colleges became eligible for federal funding, resulting in their rapid growth.

This was also a time when advances in communications and technology began to invade the core processes of colleges and universities. At first, as with most organizations, they were utilized to enable administrative, personnel, and research functions, but within the last few years, major efforts have been undertaken to employ digital and internet systems as the medium of instruction. While these developments have to date been mostly parallel to conventional educational systems, more and more colleges are experimenting with ways to combine conventional and online instruction into "blended" combinations that capture the strengths of each.

5

Diverse Colleges in Varied Sub-Regions

W. RICHARD SCOTT, ETHAN RIS, MANUELITO BIAG,
AND BERNARDO LARA

We recognize that up to this point we have been operating primarily at the macro level—about 30,000 feet in the air—to consider the composition and structure of the fields of higher education and of the regional economy and, then, the wider demographic-organizational-political-technological environment in which these systems function. In this chapter, we will bring the story down much closer to the ground.

We concluded our review of the main characteristics of the two fields by suggesting that because of substantial differences in their origins, history, structural constraints, institutional logics, and time orientation, substantial tensions existed between higher education, and the demands of the regional hi-tech economy (see table 3.1). Given these differences, how do organizations operating within these arenas connect and collaborate? This general question is best addressed by recognizing that the strength of the forces depicted vary across types of colleges and sub-regions. Research universities and comprehensive colleges, being more complex and differentiated, are better able to accommodate to conflicting pressures. Similarly, community colleges were created as hybrid organizations, educating students pursuing degrees for transfer to four-year institutions, but also to prepare vocationally oriented students for occupational careers. For-profit colleges are much less subject to the constraints of higher education control bodies and much more able to respond to market pressures.

Sub-regions also vary in industrial composition and types of personnel required. Although the San Francisco Bay Area is, relatively speaking, awash in hi-tech industries, their location and strength vary substantially across the region.

Diverse Colleges in Diverse Sub-Regions

As we began to develop our design for examining how colleges in the San Francisco Bay Area had adapted to the challenges posed by their environments, two truths loomed large: (1) colleges differ greatly in their mission—their conception of who they are and what programs they are able to provide, what supports and constraints they experience; and (2) although all operate in the Bay Area, all confront significantly different microenvironments that pose specific challenges for each. Indeed, we were surprised how much variance we found across the three sub-regions in our study.

Selecting the Sample of Colleges; Identifying the Sub-Regions

The important role of research universities in spearheading innovation and seeding economic development has been widely recognized in the media and well documented in research studies (e.g., O'Mara 2005; Saxenian 1994). These universities are of obvious significance both for the critical, direct contributions they make as well as for their symbolic role in defining what counts as "the best" in higher education. More specifically, they have provided influential models that guide the practices of other types of colleges. For example, San José State University was encouraged to devote more energy and resources to its career and vocational programs, such as engineering, by the example provided by Stanford University with its Honors Cooperative Program (see case example 3.A).

However, because so much previous media and research attention has been devoted to the research universities, particularly to Stanford and to UC Berkeley, we elected to devote primary attention to a sample of broad-access colleges. We believe that the role of these schools in helping to meet the needs of their regional political economy has not received the attention and recognition that it deserves. Broad-access colleges include a variety of public and private programs that vary greatly in size and mission. They admit the majority of students who apply and, in combination, serve the great majority of undergraduate students in this country. We decided to focus on four types of colleges—public four-year comprehensive universities, community colleges, nonprofit private colleges, and for-profit colleges, selecting at least one of each type for closer study within each of three sub-regions.

Identifying sub-regions was not a difficult task: the differences among three geographically demarcated parts of the Bay Area fairly jumped off the page.

We have already noted (chapter three) that each is serviced by its own metropolitan newspaper. Arbitrarily using county boundaries, we focused on three sub-regions:

1. *The Greater San Francisco Sub-Region* is made up of San Francisco and San Mateo Counties. These counties were home to 1.5 million residents in 2010, and they have seen the most recent rapid growth of hi-tech industries in the Bay Area during the past two decades. The area is served by the University of California, San Francisco, primarily a medical school with major research and treatment divisions. Partly due to this anchor institution, the sub-region has become an important hub of the biotech industry, hosting companies such as Genentech and Chiron. It is also the location of many social media companies, including Twitter and Yelp, and business-to-business giants like Salesforce. We include San Francisco State University in our broad-access college sample as well as the City College of San Francisco, a community college; the nonprofit college Golden Gate University; and Academy of Art University, a for-profit college (table 5.1).

2. *The South Bay Sub-Region's* boundaries are coterminous with Santa Clara County. It is the cradle and still remains the center of the Silicon Valley economy. It has grown rapidly since 1970, increasing by more than 67 percent to its present population of 1.8 million, concentrated in the city of San José. Once an area dominated by plum and apricot trees—"The Valley of Heart's Delight"—it now hosts some of Silicon Valley's most influential players, including Hewlett-Packard, Apple, Google, and Yahoo. It also is home to Wilson Sonsini, one of the most successful and influential law firms, and Kleiner Perkins, a prominent venture capital firm. It is home to Stanford University, a large and influential research university. In our sample of broad-access colleges for the South Bay sub-region, we included San José State University, Foothill and Evergreen Valley community colleges, the nonprofit Menlo College and the for-profit DeVry University.

3. *The East Bay Sub-Region* consists of Alameda and Contra Costa Counties. It was home to more than 2.6 million people in 2010. Its anchor city is Oakland, and it is home to UC Berkeley, one of the largest and most prestigious campuses of the University of California. To date, it has been less strongly involved in the hi-tech economy than the other two

Table 5.1   Case Study Schools

Region	4-Year public	2-Year public	Nonprofit	For-profit
Greater San Francisco	San Francisco State University	City College of San Francisco Skyline College	Golden Gate University	Academy of Art University
South Bay	San José State University	Foothill College Evergreen Valley College	Menlo College	DeVry University
East Bay	California State University East Bay	College of Alameda	Holy Names University	University of Phoenix

sub-regions. Instead, its industrial basis tilts toward manufacturing, which includes the solar manufacturing company Solyndra, and the personal electronics firm Logitech. Our college sample in this area included the four-year California State University, East Bay; the College of Alameda, a community college; the nonprofit Holy Names University; and the for-profit University of Phoenix. Like DeVry, Phoenix operates multiple campuses in the Bay Area and nationally, but in our sub-regional studies, we focused only on their campuses serving students in their respective areas. Of course, it is also the case that all colleges serve considerable numbers of students who do not reside in their sub-regions.

Data and Methods

While most of the research reported thus far has relied primarily on secondary data—from the Higher Education General Information Survey (HEGIS from 1970 to 1985), the Integrated Postsecondary Education System (IPEDS from 1986 to 2012), the decennial US population census from 1970 to 2010, and the scholarly work of others—this chapter and the next employ a somewhat different mix of sources. While we still rely on secondary data sources, including the above, we also utilized sources such as the Bureau of Labor Statistics Occupation and Employment Survey, the monthly Current Population Survey, and the California Employment Development Department for data on wages and industry concentration. In combination, these quantitative sources allowed us to examine changes in the demography, labor force, and market conditions in the sub-regions.

Supplementing these sources were the systematic analysis of documents obtained from the sample colleges and multiple semistructured interviews and focus groups conducted with a variety of stakeholders, including regional economists, college administrators, faculty leaders, and program directors. The sample colleges, in effect, provided a small set of cases (14), allowing us to contrast (1) differences in the activities of varying types of colleges operating in the same sub-region; and (2) differences in the ways in which the same type of college operated in one sub-region versus another. Interviews with faculty leaders and program directors were targeted to those associated with four disciplines thought to be most salient in relating to a hi-tech economy, namely, engineering, computer science, biological sciences, and business administration.

The systematic analysis of documents involved the analysis of a small sample of academic program reviews drawn from seven of the sample colleges located across the three sub-regions. The reviews focused on programs carried out in the four targeted disciplines. These reviews were first mandated by the state as a part of the assessment movement of the 1970s and 1980s, requiring faculty at public institutions to conduct, analyze, and review progress made in achieving educational objectives. They are notable in being self-studies and, because they are authored by faculty, often reflect divergence from the official policies or the goals of college administrators. We also found them to vary greatly in approach and level of detail. This data source, our research approach, and some of the findings from this work are described in more detail in a supplementary essay posted on the Gardner Center website (http://gardnercenter .stanford.edu).

More generally, to examine the structure and behavior of our sample of colleges we employed multiple methods and data sources, allowing triangulation among them (Miles and Huberman 1994). Our interviews averaged more than 60 minutes and were audio-recorded and analyzed using an iterative process of comparing and contrasting for recurring themes and patterns.

Comparing the Sub-Regions of the Bay Area

The three sub-regions vary considerably in their socioeconomic makeup. They host populations that vary in their demographics, their economic stratification, and the composition of their economies. As shown in figure 5.1, the South Bay is home to a higher proportion of Asian and Hispanic residents but fewer African Americans than the other sub-regions. Greater San Francisco

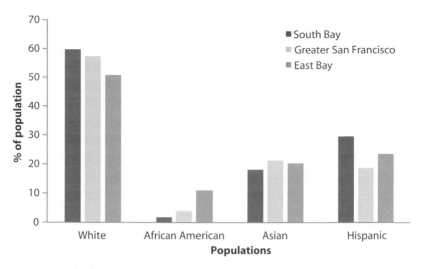

Figure 5.1. Racial/ethnic differences among the San Francisco Bay Area sub-regions, 2010. *Source*: US Census.

includes a higher proportion of white and Asian residents, and fewer Hispanics, while the East Bay is home to a higher proportion of African Americans and fewer Asians than Greater San Francisco or South Bay.

Figure 5.2 reports over time the per capita income for the three sub-regions. In 1972, there was no difference in income levels between the East and South Bays, while the Greater San Francisco area had a higher average income level. In the four succeeding decades, however, the average incomes of each geographic cluster diverged so that by 2012 there was roughly a $10,000 gap between the South and East Bay. Meanwhile, Greater San Francisco has maintained its income advantage so that by 2012, it exceeded the South Bay by more than $10,000 and the East Bay by more than $20,000. The differences in average income reflect both a difference in industrial composition as well as a difference in wages paid for the same types of occupations. For example, average wages paid for business and financial operations in 2012 were $92,000 in Greater San Francisco, $89,000 in the South Bay, and $82,000 in the East Bay; for engineering and technical occupations, they were $98,000 in Greater San Francisco, $106,000 in the South Bay, and $95,000 in the East Bay (Occupation and Employment Survey, Bureau of Labor Statistics 2012). Of course, these wage differentials also reflect differences in the cost-of-living in the sub-regions.

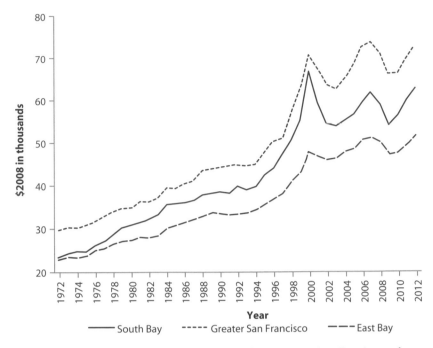

Figure 5.2. Per capita income differences among the San Francisco Bay Area subregions, 1972–2012.
Source: California Regional Economic Analysis Project, US Department of Commerce.

The industrial composition of the sub-regions also varies in important ways. The South Bay continues to be dominated by computer and electronic companies, while the Greater San Francisco area has long been a financial and transportation hub and a major tourist destination. To these specialties it has now added growing collections of biotech and social media firms. The East Bay is heavily invested in transportation: the Port of Oakland is a major shipping and transportation center. It also is home to large construction and production companies. We parse these broad differences by examining selected categories of employment within industries in the sub-regions. The categories are selected to emphasize (1) those related to the hi-tech economy, and (2) those exhibiting the largest differences among sub-regions. Table 5.2 reports differences in share of employment by sub-region for selected occupations in 2012. As expected, the Greater San Francisco area has a dispropor-

Table 5.2   Share of Employment for Selected Industries across San Francisco Bay Area Sub-Regions, 2012

Industry	Greater San Francisco (%)	South Bay (%)	East Bay (%)
Legal occupations	1.5	0.9	0.7
Business and financial operations	8.6	6.6	6.1
Leisure	12.4	9.2	9.6
Computer and mathematical	6.3	10.1	4.0
Engineering and technical	1.4	5.6	2.1
Trade (sales and transportation)	14.9	13.5	16.5
Production	2.7	5.5	5.1

Source: Occupation and Employment Survey, Bureau of Labor Statistics 2012.

tionate share of its employment in legal, business and financial, leisure, and trade industries compared to the South and East Bays. The South Bay has more of its employment base involved in computers and mathematical occupations, in engineering and technical, and in production activities; while the East Bay is more concentrated in trade, including sales and transportation.

Another area of interest concerns the location of venture capital firms, which, as discussed in chapter three, play such an important role in fueling the development of start-ups in new industries. If we compare the ranking of the key cities in this regard, we find that San Francisco is far and away the leading center of these activities, providing $10.9 billion of support in 2014 compared with San Jose with $1.1 billion, and Oakland with $200 million. If we include venture capital firms in surrounding communities, the South Bay fares somewhat better, but the rankings do not change. In 2014, the primary cities of the Greater San Francisco area provided $13.3 billion compared to the South Bay cities with $6.8 billion and the East Bay with $400 million (CB Insights: https://www.cbinsights.com/).

Common Issues across Sub-Regions
Student and Worker Mobility

It is important to acknowledge at the outset that the sub-regions are by no means isolated or self-contained entities. There is substantial mobility of students (and also faculty) across sub-regional lines. Students attending Stanford, UC San Francisco, or UC Berkeley arrive from all 50 states and from many foreign countries, and the CSU campuses attract students from across the state as well as some international students. However, geographic distance continues to have strong effects on those who attend broad-access colleges. For example, recent data for the United States compiled by Hillman (2012) shows that the median distance traveled by students of public two-year (community) colleges was 8 miles; for four-year public colleges, 18 miles; and for private nonprofit colleges, 46 miles. The relatively small distances involved reflect in part the availability of information and the importance of convenience. But, for an increasing number of students who are married, have dependents to care for, and are working part or full time, convenience can easily turn into necessity. For today's modal college student, the commute distance remains a highly salient factor in choosing a college.

There is substantial turbulence in college careers, as college students relocate from one college to another in a phenomenon known as "swirling." Data collected by the National Student Clearinghouse Research Center (Adams 2015) reveal that 37 percent of college students transfer between schools at least once within six years. Swirling is not just local; the report found that about one in five students who started at two-year public institutions and nearly one in four who enrolled first at four-year public institutions transferred across state lines. Mobility also occurs from four- to two-year colleges, most notably in the case of "summer swirlers"—students in four-year colleges who utilize community colleges for summer course work and then return to their four-year institution in the fall.

There appears to be even more volatility among students in hi-tech regions than in conventional industrial areas because of the rapid boom-and-bust cycles that sweep through the region. The movement here involves both mobility across types of schools as well as exit from and reentry to the same college. Our case studies provide anecdotal evidence regarding the ways in which these economic episodes affect student careers and college behavior.

Substantial mobility is also evident because people live and work in different places. Because of the high cost of living in the Bay Area, many workers routinely commute long distances beyond the outer limits of Alameda and Contra Costa Counties from as far away as San Joaquin and Sonoma Counties. An increasing number of the employees in hi-tech companies have recently elected to live in San Francisco to take advantage of its urban amenities and cultural zeitgeist. Companies such as Google have chartered buses, equipped with Wi-Fi, to shuttle workers between the city and their South Bay campuses. To reduce commute time, a number of companies such as Facebook, Intuit, and Apple are creating campuslike work and living complexes that encourage their employees to live, work, and play in the same place (Quinn 2014).

Variation among Companies

In their early years, during the formation of Silicon Valley, some of the larger companies were strongly embedded in the local social and political landscape. Companies such as Hewlett-Packard were literally homegrown and were not simply in but also of the community. Connections between Stanford and various corporations in the South Bay were particularly strong. As detailed in chapter three, Fred Terman, first as dean of Stanford's engineering school and later as provost of the university, was instrumental in fostering close ties between Stanford and early hi-tech companies, including Hewlett-Packard, Shockley Semiconductors, and Varian. Many of his students were responsible for start-up firms, and the creation of Stanford's Industrial Park in 1952 further fostered ties between the university and local firms. Current and future employees were trained at local colleges and universities, and firms often provided internships to students and gave equipment to schools. These early companies were closely tied to the community and engaged in wider civic and philanthropic bodies. But this began to change. One of our elite informed respondents contrasted Hewlett-Packard's earlier and later company posture:

> Bill Hewlett and Dave Packard lived in the community, were on the Palo Alto school board, for instance, and generally very involved in the community. They were engaged in the details of life in Silicon Valley. Now we have CEOs who live on an airplane and don't really live in the community anymore. [I asked a recent CEO] if it bothered him that we are producing fewer engineers in the U.S. and HP is getter fewer from the best schools. He said that as an American and a father it bothered him a lot, but as a CEO of a global company it didn't bother him at all. He could get talent elsewhere for less. CEOs have to be concerned about return on investment. . . . I see the shift in CEOs from benign patriarch in the local community to an aristocrat in the one percent of the global economy as emblematic of the change.

The largest hi-tech firms in the Valley, especially those that are publicly traded, are for the most part much less embedded in the local economy than their smaller counterparts.

Economic Fluctuations

As described in chapter three, hi-tech regions experience both high velocity change and great volatility. Not only have the types of industries dominant in Silicon Valley changed over time (see fig. 3.1), but within industries, the specific products and the versions of them are constantly being upgraded, requiring a different mix of workers and skills. Moreover, the entire Valley is subject to broader cyclical economic swings, such as the dot-com "bust" of the early 2000s or the financial crisis at the end of that decade. Colleges within the regions must attempt to accommodate to these rapid shifts. Not only do the skills they are asked to help students to acquire change periodically but also the students themselves are courted heavily during times of peak demand. At some times, as a dean at San José State University told us, recruiters are on campus offering signing bonuses, but only for students willing to start "now," not wait until June. Later, when the economy slows, many of these same students reappear in great numbers seeking reentry to the college. According to program reviews from this university, enrollment in its computer science program rose from 550 majors in 1991 to 1,100 in 2001, only to drop to 840 in 2003. The numbers never recovered and the program continued to lose enrollment over the next decade. Colleges must also deal with a scarcity of adjunct professors with industrial experience when the economy is strong, followed by a surfeit of candidates when times are slow. All colleges must find ways to cope with the challenge of rapid change.

Public colleges are beset with a second, related source of uncertainty. Because their revenues are heavily dependent on state funding, which as noted in chapter four comes from sales and income taxes, the support they receive rises and falls with the state of the economy. The ability of public universities and community colleges to meet changing demands is severely constrained by budgetary restrictions, too often resulting in cuts to popular programs at precisely the time when demand is highest (see case example 4.B).

Comparing and Contrasting Colleges in Sub-Regions
Greater San Francisco

San Francisco and San Mateo Counties are home to the Bay Area's most diversified economy. The sub-region's longstanding strengths in financial services and biotechnology have recently been joined by the northward spread of the information technology sector. In recent years, major firms like Salesforce,

Twitter, and Airbnb have made San Francisco their home, while South Bay firms such as Google operate large satellite campuses in the city. One result of this diversification is that broad-access colleges in the sub-region struggle to find a clear identity.

San Francisco State University is the area's major four-year institution, enrolling more than 25,000 undergraduate students and 4,000 graduate students in fall 2014. In-state tuition and fees totaled around $11,000 in 2014–15. The undergraduate student body is very diverse, with a roughly even split between Asian (28 percent), Hispanic (25 percent), and white (22 percent) students, followed by much smaller nonresident alien (7 percent) and African American (5 percent) student populations. By far the most popular undergraduate major is business, accounting for nearly one quarter of all students, with all other fields clustered in the single digits. The university has a six-year graduation rate of 50 percent for bachelor's degree candidates.

Interviews with SFSU administrators indicate that the university has struggled to establish strong connections to the local labor market. While some collaborations have worked, such as an internship and employment pipeline with the biotech giant Genentech, other attempts to build ties with major industries have fallen short. The university's biology department has never had an industry advisory board, while the computer science department only inaugurated one in 2005. The discrepancy between the institution and industry sometimes extends to the curriculum itself. In a notable difference from San José State, SFSU's engineering department in recent years has not pursued workforce training; a 2007 program review explained that "the target student population has shifted from part-time/currently employed students with an applied orientation to full-time students with a research orientation."

Graduate programs represent a new area of focus for the university. An administrator described the business school's MBA program as popular and profitable for the institution. In 2010, the biology department launched a "Professional Science Master's" program in biotechnology that included required cooperative internships and applied research projects. International students are also a focus of growth for SFSU, especially at the graduate level. In recent years, 33 percent of MBA students have been nonresident aliens, and the engineering department's master's programs have been majority international since the early 2000s. These students are concentrated in a few countries of origin. In 2012, over 60 percent of all applicants to the university's MS program in computer science were citizens of India; only 25 percent were Americans.

City College of San Francisco is the Bay Area's largest two-year institution, enrolling more than 23,000 students in 2014. A clear plurality of students is of Asian descent, representing more than one-third of the total, with white and Hispanic students at 23 percent each. As with other California community colleges, full enrollment is open to any student with a high school diploma or GED, and tuition and fees are low at approximately $1,500 per year.

CCSF has long been riven by internal and external challenges. Internally, tight budgets, overenrollment, and tensions between instructors and administrators are common. A 1985 biology department program review established a theme that would continue for three decades: "The faculty's efforts have been complicated by red tape, lack of funds, and an obstructionist administrative attitude. These factors tend to wear down instructor enthusiasm and to create a pessimistic apathy toward any attempts at innovation." Lack of coordination is another common problem. The college's business offerings, which represent one of its largest curricular areas, were long spread across multiple departments and campuses. The institution only consolidated them in the mid-1990s as a new School of Business, but program reviews indicate an enduring lack of overlap and coordination between programs. Interview data confirm this; high-level administrators report that many departments operate in "siloes," and that budget cuts have overburdened faculty who might otherwise work on collaborative or innovative programming.

CCSF's external challenges have come to a head in recent years with its struggle to maintain its accreditation. As discussed in more detail in case example 2.C, this dates to a 2006 review by Accrediting Commission for Community and Junior Colleges, a division of the Western Association of Schools and Colleges, which censured CCSF for budgetary mismanagement. The crisis escalated in July 2012, when the commission criticized "tangled governance structures, poor fiscal controls, and insufficient self-evaluation and reporting," and demanded immediate reform in order to avoid a withdrawal of accreditation in 2014, which would mean the closure of the institution (Asimov 2015). Recently, a state superior court ruling forestalled this action, and the college remains open. However, the crisis and ensuing publicity appear to have significantly impacted the institution's enrollment, which in fall 2014 was down more than 30 percent from just three years earlier.

In addition to these challenges, many CCSF departments have long struggled to establish strong linkages to industry. One administrator speculated that the college's social justice–oriented mission may limit collaboration with employers and workforce-oriented programming, while another noted that the institution's pay structure made it difficult to hire adjunct faculty who might have direct ties to industry. One notable exception has been CCSF's Computer Science Department, which has maintained an active industry advisory board since the early 1990s and has offered programs linked to industry-designed certificates from Microsoft, Cisco, and Oracle.

Skyline College is a much smaller community college located in San Bruno, approximately 10 miles south of San Francisco. Like CCSF, its student body, which totals just under 10,000, has a plurality of Asian students (38 percent), followed by Hispanic (30 percent) and white (20 percent). Only 3 percent of students are African American. Tuition and admissions criteria are the same as at CCSF.

More than most colleges in our study, Skyline has attempted to create formal programming linked explicitly to workforce preparation. The college's career and technical education programs are often integrated with local high schools. A particular point of pride, according to administrator interviews, is the Career Advancement Academy, launched in 2007 as a high school to workforce "bridge" program. It currently offers certificate programs in four areas (allied health, automotive, legal careers, and early childhood education), and administrators hope to launch a biotechnology track soon.

Like other community colleges, however, Skyline struggles with budget problems and uncertainty about its identity. One administrator noted the sharp contrast between the college administration's desire to focus on career education and the demands of students, many of whom say they are focused on transferring to four-year institutions. The college has significant success in this area, ranking as one of the top three California community colleges in terms of its transfer rate.

Golden Gate University is a broad-access nonprofit university headquartered in downtown San Francisco. As a standalone institution, it dates to 1923, and currently serves a student body dominated by graduate students. Today, only 17 percent (489) of its students are undergraduates, but this number has fallen dramatically over the past two decades; in 1996, it enrolled nearly four times as many students in bachelor's degree programs. Undergraduate enrollment dropped off dramatically after the 2000 dot-com crash and has never recovered.

A plurality of undergraduate students is white (27 percent), with Asian and Hispanic students representing 15 percent each. Golden Gate has the largest percentage of African American students (12 percent) of any school we studied in the greater San Francisco sub-region. Graduate students are distributed more unevenly; 31 percent are white, followed by Asian (17 percent) and non-resident alien (16 percent), with much lower numbers of Hispanic and African American students.

Undergraduate tuition and fees at Golden Gate totaled $21,370 in 2014–15. While this is higher than the public institutions in our study, it is much lower than many fellow nonprofits in the Bay Area. The university does not make its acceptance rate public. The university's undergraduate programs are all related to business operations and services, including majors in accounting, human resources, international business, and supply chain management.

Golden Gate's much larger graduate division replicates all of these business programs, and adds two master's degree programs in psychology and a law school that enrolled 301 JD students in 2014. One of Golden Gate's strongest disciplinary areas is taxation, including seven master's degree and certificate programs in tax law, accounting, and financial planning. In the 1970s, it opened satellite campuses in Seattle and Los Angeles devoted to taxation programs. These continue today and are closely integrated with industry (the Los Angeles branch is located within the regional headquarters of the tax accounting firm Ernst & Young). The university also has a satellite campus in the South Bay city of Santa Clara, known as "GGU Silicon Valley." It primarily offers evening and weekend classes to working students. Golden Gate offers many of its courses and programs online, although it does not provide statistics on the number of students enrolled in that format. An administrator reported that 85 percent of the university's curriculum is delivered by adjuncts, many of whom are active practitioners in business services.

Academy of Art University is a large, for-profit, broad-access institution spread across several San Francisco campuses. It is classified as a special-focus institution since the vast majority of its course work is related to one disciplinary area (visual art). In 2014, it enrolled just over 15,000 students, of whom two-thirds were undergraduates. Its biggest demographic by far is nonresident aliens; 34 percent of all students are citizens of foreign countries. White students make up the next largest group (23 percent), followed by much smaller numbers of Hispanics (9 percent), Asians (7 percent), and African Americans (7 percent).

Academy of Art does not report admissions criteria, but its degree programs are largely open to any student with a high school diploma or GED. Its six-year graduation rate for bachelor's candidates is 31 percent, significantly below that of SFSU (the only other institution in the sub-region to report this statistic). Tuition and fees for full-time undergraduates total $25,750. The institution has experienced fluctuations in its enrollment over the past two decades. In 1996, it enrolled just 2,675 total students. This number rose steadily to 18,093 in 2011, but has since fallen 16 percent to its current level. This may be related to the overall downward trend in for-profit college enrollment, discussed in case example 5.A.

CASE EXAMPLE 5.A
CHALLENGES FACING FOR-PROFIT COLLEGES

Internal Practices

As argued by Tierney and Hentschke (2007: 1), for-profit universities "represent a new, fundamentally distinct type of postsecondary institution." They point to a number of factors that distinguish them from traditional colleges and universities. The first is a minimalist, highly vocational curriculum, conceived and developed not by faculty members but by administrators in the institutions' corporate headquarters (Tierney and Hentschke 2007: 94–96). This approach allows for quick adaptation, but it also creates narrow credentials that have questionable long-term value in a rapidly evolving labor market like Silicon Valley's. The second attribute is the nature of for-profit institutions' faculties, which are largely adjunct and exclusively teaching-focused. Additionally, faculty members have limited autonomy; course loads, curricula, and assessments are tightly controlled by the institution (Tierney and Hentschke 2007: 100–106).

Performance Issues Resulting in External Scrutiny

The largest challenge facing for-profit colleges in recent years has come from the tendency of some colleges to engage in fraudulent practices that have occasioned increased scrutiny by the federal government. The impetus for the investigation was the enormous federal investment in the sector; in 2009–10, $32 billion in federal student loans and grants flowed to for-profits, accounting for nearly all of the institutions' revenue stream. For-profits received 25 percent of all federal student aid given in that academic year, although less than 13 percent of students nationwide attended such an institution (Harkin 2012: 12–14).

A congressional investigation found fault with dozens of for-profit institutions for aggressive and deceptive recruiting practices, excessive tuition, and poor student support structures. The committee identified three major adverse consequences for students. The first was high debt; 57 percent of for-profit students left college (either by graduating or dropping out) with over $30,000 in debt, compared to 25 percent of students at private nonprofits and 12 percent from public institutions (Harkin 2012: 130). The second was high unemployment. Despite a heavy vocational focus in the curricula of for-profit colleges, in 2012 23 percent of their former students were unemployed and looking for work (Harkin 2012: 138). These consequences combined to cause a third concern: very low loan repayment rates and the specter of widespread defaults. In 2012, 64 percent of former for-profit students were either not making payments on their loans or were only making interest payments; 22 percent had defaulted within three years of leaving school. The Department of Education estimated that 46 percent of for-profit students would eventually default (Harkin 2012: 131–33). This situation leads to both ruined credit for borrowers and major liability for taxpayers.

After years of litigation, a federal court finally affirmed the legality of a revised gainful employment rule, which went into effect in July 2015. The new version targeted academic programs, not institutions, and focused on debt-to-earnings ratios of former students, rather than default rates. The Department of Education estimated that 1,400 programs would become ineligible for funding within a year of the rule's implementation; 99 percent of these were at for-profit colleges (Lough 2015).

The federal government was not the only source of external scrutiny of for-profit colleges. In 2013, California Attorney General Kamala Harris filed suit against Corinthian Colleges, Inc., a for-profit company that owned subsidiaries operating as Heald College, Everest College, and WyoTech College. She argued that the "predatory" colleges "intentionally targeted low-income, vulnerable Californians through deceptive and false advertisements and aggressive marketing campaigns that misrepresented job placement rates and school programs" (Office of the Attorney General 2013). In the face of the lawsuit, an investigation by the federal Consumer Financial Protection Bureau, and a $30 million fine from the US Department of Education, Corinthian crumbled, leading to its bankruptcy and closure of all campuses in 2015. The company operated nine campuses in the Bay Area, spread out among all three sub-regions in this study. They collectively served 14,000 students in the area, representing nearly 20 percent of Corinthian's nationwide student population (Murphy 2014).

Enrollment Challenges

For-profit colleges also face unique challenges with their enrollment numbers. While their flexible staffing and heavy use of online instruction allow them to rapidly expand capacity in response to market demands, they are also vulnerable to shifts in student preference. After a decade of explosive growth, enrollments in for-profits began to fall precipitously in 2010. Since then, the total number of students at degree-granting four-year nonprofits has fallen 25 percent (Smith 2015). Some institutions have seen much greater losses, most notably the University of Phoenix, which lost more than 50 percent of its students during that period and has predicted continued drops totaling to 70 percent of its peak enrollment (Gillespie 2015; Hansen 2015).

Some of this decline in enrollment is undoubtedly due to negative publicity related to governmental scrutiny of for-profit institutions and increased media attention on the sector. Additionally, some colleges have faced strenuous public criticism from their own students, some of whom are refusing to pay their loans on principle, claiming that they received deceptive information and an inferior education (Lewin 2015). Finally, the sector is facing increasing competition from public and nonprofit institutions that have replicated some of its practices (especially online course work) at much lower prices.

The institution was founded in 1929 with strong ties to the advertising industry. While advertising art continues to be offered, Academy of Art now offers many programs spread across 23 areas ranging from art history to fashion. While the institution's website boasts internship opportunities in many fields, we were unable to find clear ties to industry in either interviews or document analysis. According to the DOE's "College Scorecard" tool, 52 percent of Academy of Art graduates earned more than the average high school diploma holder, compared to 66 percent at the much less expensive SFSU.

East Bay

The East Bay is already the Bay Area's largest sub-region, and estimates predict that the population in Alameda and Contra Costa counties could climb as high as 3 million by 2020. Unlike the thriving hi-tech economies of the South Bay and Greater San Francisco area, industry sectors in the East Bay have been slower to recover from the Great Recession. Sectors predicted to see the most gains include transportation and warehousing (4.9 percent), real estate (4.2 percent), and wholesale trade (4.0 percent)—largely as a result of increased activity in the Port of Oakland (http://eastbayeda.org/ebeda-assets/reports /2014/EDA-Outlook-2014-2015.pdf). Workers in high-skilled occupations, such as information and professional, scientific, and technical services, are more likely to commute from the East Bay to San Francisco and the South Bay, where a higher concentration of such jobs are located.

Many postsecondary institutions serve students in the East Bay, including California State University East Bay; a host of for-profit institutions focused on technology, business, and health care; and ten community colleges. The present study focused on CSUEB, the two-year College of Alameda, and two proprietary institutions: the University of Phoenix and DeVry University. We also interviewed leadership from the Peralta Community College District, which constitutes the College of Alameda, Berkeley City, Laney, and Merritt Colleges. Interviews with faculty, administrators, and other leaders suggest that the East Bay is largely removed from the rapid technological growth and economic development seen in the South Bay and Greater San Francisco area.

Founded in 1957, the *California State University East Bay* (CSUEB) is a public institution that enrolls about 15,000 students, a majority of whom are undergraduates. Roughly 25 percent of students identify as Hispanic, 22 percent as Asian, and 10 percent as African American. Given its large enrollment of racial/ethnic minorities, in 2011 the federal government designated CSUEB as an Asian and Pacific Islander Serving Institution (APISI). In April 2014, the university was also designated as a Hispanic Serving Institution (HSI).

CSUEB's main campus is located in the city of Hayward, the sixth largest city in the Bay Area with a population of roughly 150,000 people. CSUEB's second campus is located in the Concord foothills of Mount Diablo, and it offers bachelor's degrees in fields such as criminal justice administration, nursing, and sociology. The university also has a center for professional development located in downtown Oakland that offers continuing education (e.g., state notary public training) and certificate programs (e.g., pharmacy technician); the center also provides corporate training facilities such as flexible meeting rooms, wireless laptops, and video conferencing equipment.

In 2014–15, CSUEB accepted about 70 percent of the students who applied and charged in-state students roughly $6,500 for tuition, with out-of-state students paying significantly more at $17,724. Approximately 76 percent of resident, full-time, degree; and certificate-seeking undergraduates at CSUEB received some type of financial support (e.g., Pell Grants, federal student loans, institutional scholarships).

CSUEB offers a variety of undergraduate majors, with the bulk of its students completing degrees in business administration, health sciences, nursing, psychology, and criminal justice administration. In fact, business and nursing are the college's two impacted fields of study; no other areas are oversubscribed. Leading graduate majors include social work, educational leadership, and business administration. Six-year graduation rates for those seeking bachelor's degrees are roughly 41 percent, about six percentage points lower than San José State University.

In the mid-2000s, CSUEB suffered from low enrollment and, as such, devised ways to attract more students such as implementing aggressive recruitment activities in the central and southern portions of California. The university also built dining halls and dormitories, positioning itself as both a destination and commuter campus. Administrators identified active recruitment of international students from China, India, and Saudi Arabia, among other countries, as a key strategy in boosting enrollment and enhancing revenue.

CSUEB also developed programs that were not provided elsewhere in the Bay Area to attract students, including new training schemes in civil engineering and construction management. Further, the University expanded its course offerings in biochemistry to meet regional labor market demands. One administrator stressed how it is critical for the university to hire faculty that have an "entrepreneurial spirit" and can predict the fields of study and types of competencies students will need in the near future.

Yet establishing new programs in the CSU system is a highly complex and lengthy process. Top administrators at CSUEB report how bureaucracies in the current system (e.g., course approval processes) hinder their ability to effectively respond to students' needs and demands. Additionally, years of persistent cuts in financial support and funding models that are directly tied to enrollment make it challenging to establish and sustain programs, particularly career and technical education (CTE) course offerings in STEM-related fields (e.g., biotechnology) that frequently require high-cost laboratories and equipment.

Administrators from the *College of Alameda*, a small two-year public community college, similarly noted how CTE programs were cost prohibitive. Unlike CSUEB, however, courses in automotive technology, aviation maintenance, and transportation and logistics support were more popular than those related to biology or computer science. The College of Alameda (which is also discussed in case example 5.B) opened in 1968 and is one of the four colleges affiliated with the Peralta Community College district. Located on Alameda and Bay Farm Islands, the suburban city of Alameda is home to an estimated 78,000 people, of which more than half identify as white (57 percent), followed by Asian (36 percent), with smaller percentages of Hispanics and African Americans.

CASE EXAMPLE 5.B
THE FOOTHILL-DE ANZA AND PERALTA COMMUNITY COLLEGE DISTRICTS

The high-velocity labor market of the San Francisco Bay Area presents unique conditions and challenges to the region's community colleges. Our data suggest that colleges, as well as entire districts, vary in their response to particular regional needs. For instance, we observed variations in colleges' connections with hi-tech companies, the types of students they serve, and the level of resources available to create responsive programs. To illuminate some of the differences among colleges in our sub-regions, we compare the Foothill-De Anza and Peralta Community College Districts.

The Foothill-De Anza Community College District
As one of the largest community college districts in the country, Foothill-De Anza is located in the South Bay—the heart of Silicon Valley. Data from fall 2014 indicate that the Foothill-De Anza district served about 38,000 students, a majority of whom identified as Asian (30 percent), followed by whites (27 percent), and Hispanics (25 percent).

Enrollment among Asian and Hispanic students at Foothill College has more than doubled since the 1990s, which reflects the increasing population diversity in San Mateo and Santa Clara counties. In addition, the district has the second largest international student population of all two-year institutions nationwide. Data indicate that the number of students 25 and over have declined over time. Administrators comprise roughly 3 percent of the district's staff, while 26 percent were tenured faculty, 52 percent were temporary employees (e.g., adjunct lecturers), and 21 percent were classified employees. Foothill and De Anza Colleges have ranked among the top two-year colleges in the country. The district is known as a pioneer in online education and in offering computer and technology-related training schemes. Foothill College, for instance, offers a wide variety of online classes that can help satisfy associate degree requirements in areas such as accounting, graphic and interactive design, economics, and music technology (Foothill College website). The college also offers a range of certificates (e.g., certificates of achievement, career certificates) whose courses can be taken entirely online; among other areas, students can earn a certificate of proficiency in bookkeeping, payroll preparation, and as a tax specialist.

Similarly, De Anza College also provides online or distance learning opportunities in areas such as business law, managerial and financial accounting, and entrepreneurship. Moreover, since 1984, De Anza College has had a Department of Computer Aided Design (CAD) in the Business/Computer Systems Division, whose aim is to "meet the changing needs of the local Silicon Valley engineering and design firms" (De Anza College website). The Business/ Computer Systems Division offers certificate and degree programs in computer information systems, computer applications and office systems, and manufacturing and computer numerical control (CNC) technology, which trains students to become CNC machinists and programmers as well as industrial engineering technicians—job areas estimated to grow in California.

In addition to its traditional course offerings, in 2013 the Foothill-De Anza district won a $16 million grant to establish a new statewide online program aimed at increasing access to online classes, with the hopes of boosting the number of associate degrees and students transferring to four-year universities. To match students with skills needed by local employers, the district is currently building a new $25 million 50,000 square-foot education center in Sunnyvale's Moffett Park, which will include training and curriculum in high-demand fields, including geographic information systems (GIS) technology. This new center will be located near many hi-tech companies such as Ruckus Wireless and Google.

Differences in Establishing Partnerships

While both Foothill and De Anza Colleges are situated near Silicon Valley and strive to offer market-responsive course offerings (both in person and online), our interviews suggest that they differ in their ability to partner with local companies. A former administrator at De Anza College, for instance, expressed his struggles: "The large companies are indifferent . . . we're not important to them. . . . Apple, HP, Cisco, we're lucky if we get a few internships . . . there is very little connection. Their view is that nothing in the infrastructure of the Valley is ramped up enough to give them the specialized expertise and knowledge and address what they need."

In contrast, a former administrator at Foothill College, which is only seven miles away, was able to strike partnerships more successfully. She explained how the college was able to secure "a lot of programs with Intel, Hewlett-Packard, and other companies." She recounted one partnership with Tandem Computers: "In that negotiation, we helped rewrite the manuals and we figured that they (Tandem) needed technicians. So we asked them to give us their equivalent course work, and we offered to train workers for them; both Tandem employees and potentials. We'd give them certificates. . . . This connected folks together; connect students to employers and to see whether there was a match." This former Foothill College administrator explained that leadership was key in establishing these ties to industry, particularly since programs were (and continue to be) quite costly to run: "You need coherent leadership in the college, in the board, in the chancellor, and on to and through the faculty. The president needs to raise the stature and level of importance of the career tech programs and faculty; that's number 1; and number 2, the people need to go out and make those industry contacts."

The Peralta Community College District

The Peralta Community College District (PCCD) serves northern Alameda County in the East Bay, one of the most culturally diverse counties in the country. PCCD operates four colleges: Berkeley City College, Laney College and Merritt College in Oakland, and the College of Alameda.

Today, the Peralta Colleges offer degrees and certificates in roughly 50 different technical fields, including advanced manufacturing, construction, global trade and logistics, advanced transportation, as well as information and communication technologies (Peralta district website). However, currently, the Accrediting Commission for Community and Junior Colleges has placed the College of Alameda and Merritt College on probation, and issued warning status for Laney College and Berkeley City College. The commission found that PCCD's colleges were deficient in, among other things, developing academic supports for distance learning students, ensuring fair allocation of resources, and adequately monitoring students' achievement. The commission has given the colleges until October 2016 to resolve these and other issues (White 2015).

According to a former administrator of the Peralta Community College District, the colleges have distinct identities and missions, and the district's structure makes it difficult to understand who has leadership authority. He explained, "Merritt College, which used to be the Oakland City College, was home of the African American Panthers. . . . Merritt College has all the allied health programs. Laney College started in the 1970s, started as a vocational school. . . . Alameda has the problem of still defining its mission. Berkeley City sees itself as a transfer college; hi-tech emerging there. . . . The challenge is 4 colleges and 1 chancellor, 1 district office. There is some ambiguity in terms of who's in charge."

Data from fall 2014 indicate that the Peralta Community College District served about 28,000 students, 10,000 students less than Foothill-De Anza, with most students identifying as African American (24 percent), followed closely by Hispanics (23 percent) and whites (20 percent). PCCD's colleges serve distinct populations; for instance, Berkeley City College serves mostly white students, while African American students make up a majority of those enrolled in Merritt College. The College of Alameda, in contrast, is seeing a surge in the number of Asian students—increasing from 17 percent in 1990 to 36 percent in 2010. In terms of staff, data indicate that about 3 percent of PCCD hold administrative positions and 23 percent are tenured or tenure-track faculty. Temporary employees make up about 57 percent of the staff while classified employees make up 18 percent.

Colleges in the East Bay district also vary from one another in the types of courses they offer students. For instance, while Berkeley City College provides a host of courses in computer information systems (e.g., network support technician, web scripting, advanced computer programming), the College of Alameda, which is only eight miles away, offers training in automotive technology; transportation, distribution, and logistics; warehouse and forklift operations; and diesel mechanics. In fact, the College of Alameda's ATLAS program (Alameda Transportation and Logistics Academic Support), part of the California Transportation and Logistics Initiative (CATLI), is a direct response to the job needs stemming from the Port of Oakland about three miles away.

An Absence of Hi-Tech Partnerships

Unlike the Foothill-De Anza Community College District, the Peralta Community College District reports more obstacles in establishing links with industry players, in part due to their location, which is roughly 40 miles northeast of most Silicon Valley companies. As one district administrator put it, "I think (San Francisco) City College and University of San Francisco are benefiting hugely from Salesforce and these guys really saying Mark Zuckerberg's giving all this money . . . Oracle, Google . . . Great, awesome. They're finally realizing that this is an important piece of their ecosystem. We don't have any of those people here, in the East Bay. Rarely does their view reach across the bay to us." For example, when asked to describe the student body, one district administrator said, "Some want programs that are quick, easy, and will get them a job quickly. Some were formerly incarcerated. Their lack of computer skills impedes their job searches. . . . I've noticed a lot of CTE students are not very computer savvy." A respondent from the College of Alameda expressed, "We are more of a CTE college. . . . Here at the College of Alameda, it's all about transportation logistics because of the Port of Oakland. The old army base in Oakland is being transformed into a logistics hub; it has the potential to be a huge employer." The same administrator explained, "It's hard for us to find classrooms. For our ATLAS program this fall, we could not find space for two of our classes. Fortunately, our Bay Ship and Yachts partner offered their space and their workers can sit in on the classes. The space issue is going to get worse." One administrator remarked, "Most of our full-time faculty—we call them CTE people—have to have a minimum of five years of industry training, but maybe theirs was 20 years ago. We don't necessarily have either people or structures that keep them well connected."

The challenges we observed in the College of Alameda partially stem from structure-related obstacles, including the time it takes to approve new courses within the community college system. When asked to comment about the quality of the district's business course offerings, one respondent described, "It's almost embarrassing. Part of this is because of the governance structure of creating new programs and classes. There is an inability to have a teacher be nimble, to incentivize, to be ahead of the game. . . . There's some innovation but business courses are basically, unfortunately, bad and not that innovative."

Similar to other institutions in our study, years of higher education budget cuts have limited the PCCD's capacity and inclination to partner and innovate.

One administrator put it this way: "For years, we've been chronically under-funded. . . . A lot of the long-time faculty members are very turf-oriented or entrenched . . . they fought hard for their program, and they are not going to give an inch. It takes a certain level of trust. . . . We're just a much more economically hard pressed region that, unfortunately, can lead to a region that is not as willing to take risks . . . or collaborate."

In summary, evidence indicates that community colleges in the San Francisco Bay Area vary in their responsiveness to local labor market needs, depending on—among other things—local conditions, students' needs, and physical capacity. In consonance with the literature (e.g., MacAllum and Yoder 2004), we find that issues related to resources and funding, leadership and governance, and organizational structure bear upon the institution's ability to be market responsive.

More ethnically diverse than the city it serves, the College of Alameda enrolls a population of roughly 6,000 undergraduate students, 30 percent of whom identify as Asian, 22 percent Hispanic, and 22 percent as African American. Over 40 percent of the student body are aged 25 or older, and a majority are California residents paying an in-state tuition of about $1,100; nearly 80 percent of students receive some type of grant or scholarship aid.

The College of Alameda offers a range of associate degrees (e.g., accounting, dental assistant), certificates of achievement (e.g., engine repair specialist, diesel mechanics), and certificates of proficiency (e.g., small business administration, warehouse and forklift operations). It also offers associate transfer degrees in fields such as business administration, math, and psychology, along with services and workshop events that educate students—many of whom are the first in their families to go to college—about transfer requirements to the CSU or UC system, the financial aid process, and housing.

In contrast to its counterparts in the Foothill-De Anza Community College District, the College of Alameda serves a larger share of low-income, first-generation, and ethnic/minority students (this contrast is discussed in case example 5.B). Respondents observe how most of their students are intent on finding or advancing their careers by obtaining in-demand technical skills and participating in meaningful apprenticeship programs that lead to viable jobs. Given the college's close proximity to the Port of Oakland, disciplines such as transportation, distribution, and logistics are popular in the college.

However, courses are difficult to offer since there are few on-campus facilities that can house this type of training, particularly as the campus undergoes renovation and classrooms are relocated to temporary bungalows and other shared spaces. Administrators in this sub-region argue that there is ever-growing investment in facilities that support technological innovations in information and computer science and biology, and decreasing support for skill development in manufacturing or transportation and logistics. One East Bay leader remarked, "Colleges go for the low-hanging fruit. It's cheaper to produce a business degree than someone with technical skills."

Moreover, respondents argue that there are significantly fewer technology companies doing business in the East Bay. Although major companies such as Bayer, Kaiser, Clorox, and Safeway have had a long history in this sub-region, the growth of new companies and start-ups has been modest relative to the South Bay and Greater San Francisco area. Our interviews suggest that industry partnerships are much fewer and more competitive in this sub-region, particularly during economic downturns. Additionally, according to one administrator, the interests and investments of local groups such as the Bay Area Council and Joint Venture Silicon Valley are localized, with few expanding outside of their immediate area.

Partnerships to support students' career readiness are also rare among the colleges themselves. Given diminishing resources for higher education, colleges in the East Bay are often in competition for students and, as such, have to be strategic in the types of programs and courses they offer. One respondent remarked, "They don't want duplication of programs. We have to keep tabs on service areas and make sure that we don't over draw from each other. Each college gets allocations based on their enrollment." Another said, "Even though programs in San Francisco and San José are impacted, we can't claim those students because of a 'do-not-compete' clause between the colleges. For example, we're not supposed to recruit in San Francisco, Hayward, or Fremont."

Administrators at *Holy Names University*, a private, nonprofit, coeducational university in Oakland, also reported difficulty in establishing community and industry partnerships. Affiliated with the Catholic Church and administered by the Sisters of the Holy Names of Jesus and Mary, Holy Names was founded in 1868 and currently enrolls about 1,200 (mostly in-state) students, of whom 26 percent identify as Hispanic, 25 percent as white, and 22 percent as African American. Recent figures indicate that it costs undergraduates about

$35,000 per year to attend Holy Names, with 80 percent receiving some type of grant or scholarship aid.

The university offers certificate programs (e.g., educational therapy) and undergraduate training in business, accounting, biological science, and multimedia arts and communication, among others. In addition, Holy Names offers a nursing program, graduate training in a variety of fields including business administration, education, and counseling, as well as education credential programs (e.g., multiple subject credentials). Overall graduation rate among undergraduates is 50 percent, with most completing degrees in health professions and related programs; business, management, marketing, and related support services; and psychology. Unlike the College of Alameda, Holy Names does not focus on preparing students for careers in transportation and logistics despite its proximity to the Port of Oakland. Yet similar to most East Bay community colleges, Holy Names has struggled with sustaining business partnerships and enrollment. Administrators also report a significant shift over time in the type of students that enroll in their institution, in that students are older, much more ethnically diverse, and choose to attend classes online or in the evening due to part-time work commitments.

The *University of Phoenix*'s Bay Area campuses, with locations in Oakland, Livermore, and San José and a student count of roughly 5,000 (in 2015), also enroll many students that are considered "nontraditional" (e.g., first-generation students, single parents). Data obtained directly from the university indicate that since 2010, enrollment has declined for all age, race/ethnicity, and immigrant status groups. In 2014, undergraduate students were paying an average of $12,500 in tuition and fees, while graduate students were paying $17,000; about 70 percent of students received some form of financial aid (e.g., Pell Grants, veteran's benefits, employer paid tuition benefits). As a for-profit entity, tuition and fees are by far the leading sources of revenue for the university. An early administrator observed that in the 1990s, more than half of tuition revenue came from companies in the form of tuition reimbursement for their workers, and that this has since changed with the "dot-com bust," with more workers feeling the need to get retrained for the new economy.

Although students can take courses online or in person (or some combination of both), most Bay Area University of Phoenix students exclusively pursue their degrees online (80 percent) versus attending a physical campus. The share of Hispanic students taking classes through online platforms has grown from

6 percent in 2000 to 15 percent in 2015, while the proportion of white students has declined from 29 percent to 16 percent. Still, regardless of the instructional type and the various kinds of supports available to students (e.g., academic and financial aid advisors, success coaches), overall graduation rates remain low at 20 percent even though about 80 percent come to the institution with some prior educational credits or "life experience." As a result, the University of Phoenix in the past several years has and continues to experience scrutiny from consumers and the federal government alike (as discussed in case example 5.A).

The University of Phoenix prides itself on its close partnerships with local and national industries (e.g., Chevron, Pacific Gas and Electric [PGE]) and being able to offer undergraduate and graduate courses in a variety of technical fields, such as technology, business and management, and nursing and health care. Respondents noted how the use of industry advisory boards helps develop and maintain contemporary and market-responsive training programs. By convening Bay Area associations, businesses, and companies, the university is better able to identify the core competencies and skills needed in today's labor market. Further, respondents from the university also reported being able to offer training schemes to employers that help meet the continuing educational needs of their employees; for instance, locally, they help Cisco in training their engineers to be better managers.

Unlike the community college system, the University of Phoenix's centralized, top-down model allows the institution to be nimbler in building upon or modifying existing course work or creating new programs. This centralized model also allows for greater flexibility in the hiring, training, and firing of instructors. One administrator observed how accreditation standards, which are used by public institutions as a quality measure, are perceived by many in the University of Phoenix as "bare minimum and not really helping to improve programs."

South Bay

We have previously pointed out that the South Bay was the prime locus for the generative activities that developed into the Silicon Valley phenomenon. This sub-region is host to many mature software development firms, specialists in legal issues, and temporary employment agencies that focus on the software and technology sector. We would expect that the broad-access colleges located in this area would be substantially affected by and active participants in this process. For the most part, our in-depth studies indicate that this was the case.

San José State University (SJSU) provides a strong example of a college that has both benefited from and contributed to the success of Silicon Valley. Indeed, their unofficial motto is "Powering Silicon Valley." The university served more than 26,000 undergraduate and 6,000 graduate students in 2014. Tuition for in-state undergraduates was $7,300, with fees adding $2,000; for out-of-state students, $18,500. In 2014, a plurality of students was white (43 percent), followed by Asian (34 percent), Hispanic (20 percent) and African American (3 percent). In terms of bachelor's and master's degrees awarded, the five largest programs at the university were business (1,400 graduating students), engineering (1,100, of which 57 percent were MS degrees), health professions (640), visual and performing arts (480), and education (460). SJSU claims to be the number one supplier of education, engineering, computer science, and business graduates to Silicon Valley (see SJSU website). Indeed, numerous authorities agree that San José State has trained more of the engineers in the Valley workforce than either Stanford or UC Berkeley.

As detailed in case example 5.C, SJSU's College of Engineering has adapted well to the changing needs of the Valley. In its early years, it had close ties to the area's aerospace industry; later its emphasis shifted to training both hardware and software engineers for the emerging information technology sector. According to engineering leaders, for a time SJSU attempted to map changes in the economy by developing entirely new majors. For example, 20 years ago, the College created a new curriculum for majors in semiconductor technology, but by the time it was ready, the industry had begun to move offshore and demand had dwindled. As a consequence, the College now offers a basic engineering program but changes the instructional materials and labs within the program to keep abreast of current trends in the Valley. Basic curricular changes are also constrained both by ABET, the accreditation body for engineering and computer science, and by the CSU system, which must approve all new course offerings. In any case, trying to keep up with the latest, newest thing may not be a good long-term strategy. A faculty leader pointed out that to be successful in Silicon Valley, students do not simply need the latest programming or technical skills to fit them for employment; rather, she explained, "Our students need to be able to be lifelong learners. This is really important for San José State. We want them to keep going beyond that first job. Student success is not measured during their time in college, but by where they enter the workforce and how they transform their career."

CASE EXAMPLE 5.C
SAN JOSÉ STATE UNIVERSITY SCHOOL OF ENGINEERING

Today, the School of Engineering at San José State University (SJSU) enrolls nearly 10 percent of the university's undergraduate student body. Engineering course work at the institution dates to 1946, and in its early years the department was known for its specialty in aeronautics and maintained close ties to industry leaders, including Lockheed. In 1967, the department became the School of Engineering, complete with its own dean (Hasegawa 1992: 78–96).

Despite a boost in prestige when its parent institution earned university status in 1972, the School of Engineering went through a stagnant period in the late 1960s and early 1970s. Due to the wind-down of the Vietnam War and the energy crisis of 1973, the aerospace industry in California shrank considerably. By 1978, the school was once again adding new programs and enrollment reached an all-time high of 2,360 students (Hasegawa 1992: 109–25).

Much of this change in fortune can be attributed to the information technology industry, which began its explosion in the Bay Area in the 1970s. Initially, the San José State engineering programs were not set up to feed a human capital pipeline into these cutting-edge technologies. However, the booming Silicon Valley economy called for major infrastructure development led by civil engineers and construction experts, which the School of Engineering was now producing in large numbers (Hasegawa 1992: vii).

At the beginning of the 1980s, the School of Engineering was still limited by lack of funds. Due to a shortage of space, only 33 percent of applicants were accepted, and the school faced $20 million in anticipated costs to replace outdated equipment. One way to mitigate the crisis was a move away from full-time tenured faculty. In 1979, 40 percent of instruction was already taught by part-time instructors, many of whom also worked in industry. Building on these connections, the school's new dean, Jay Pinson, arranged for faculty members to take year-long sabbaticals to work in industry. These connections yielded good results for students: during the 1980s San José State supplied more than twice as many engineers to Silicon Valley firms as any other single university.

Close ties to industry paid off in other ways as well. Early in the decade, the school established an Industry Advisory Council that helped identify long-term goals and develop curriculum. It also literally helped build the school. Project 88, an ambitious renovation and expansion of the School of Engineering, was the largest capital project ever pursued in the CSU system at the time of its 1985 launch. Its end-goal was to allow the school's total enrollment to grow to 4,255 by 1990. Two-thirds of the funds came from the state of California, but the $13 million balance came in direct contributions from firms with close ties to Silicon Valley, including Lockheed, IBM, and Hewlett-Packard (Hasegawa 1992: 129–42).

As Silicon Valley became less defined by manufacturing, SJSU adapted, inaugurating a computer science program in 1976 and a computer engineering program in 1990. Early in its history, it was oriented toward the energy industry in the East Bay sub-region, then shifted toward the semiconductor industry, and in recent years has reoriented toward the pharmaceuticals industry. While the School rarely inaugurates a formal new major, the administrator reported that faculty have significant leeway to "change the focus within the majors. . . . We can bring in new technologies as well and just change the labs." In addition to continuing work with industry advisory boards (one for each of the ten departments), this leader reported that the School's ongoing reliance on adjunct instructors helps reinforce industry connections, citing both currently employed engineers who teach in the evenings and retired engineers interested in teaching as a second career.

In 2001, the Computer Engineering Department, less than a decade old, had grown dramatically. It enrolled 1,433 undergraduate majors and employed 33 faculty members, 60 percent of them adjuncts. Three years later, however, following the dot-com crash, enrollments were down to 690 undergraduate majors. The numbers have not since recovered and today are even lower than in 2004. However, close ties to industry buoyed the department, with gifts of equipment and direct grants that averaged $1.5 million per year during the early 2000s, as well as an industry scholarship program that fully funded 25 computer engineering students between 2001 and 2005.

A program review completed in late 2006 noted "a radical shift from undergraduate to graduate enrollment." Indeed, graduate students have continued their prominence in the department; in 2012–13, the department awarded 84 master's degrees in computer engineering versus 52 bachelor's degrees. The Software Engineering Department awarded 181 master's degrees versus just 13 bachelor's degrees. While the SJSU computer engineering program faces competition from doctoral-level institutions like Stanford and Berkeley, it is almost unique among CSUs. Only one other campus (Sacramento) awarded any master's degrees in the discipline in 2012–13, and it only awarded six, compared to SJSU's 84. For software engineering, SJSU awarded 79 percent of all CSU master's degrees and 62 percent of all CSU bachelor's degrees. (For comparison, across all disciplines, it accounted for 12 percent of all general CSU master's degrees and 6 percent of all general bachelor's degrees.)

More senior college administrators have also played important roles in keeping the university abreast of developing needs in the Valley. For example, the Office of the President and the institution's provost have encouraged the formation of an interdisciplinary faculty learning community to develop new programs clustered around cybersecurity and the exploitation of big data. SJSU has also experimented with the use of online learning. For example, in 2012 they collaborated with the developers of the edX platform to utilize their programming for a course on electronic circuits, blending them with in-class discussions and exercises. The university has also offered Udacity courses as components of its SJSU Plus program targeting at-risk and underserved students (Walker 2013), although that effort has had limited success, as discussed in case example 4.C.

In a manner parallel to San José State, *Foothill Community College* has exhibited substantial entrepreneurial energy in adapting to the needs of the Valley economy (see case example 5.B). In 2014, Foothill enrolled 15,000 students, a third of whom were full time. The entering class included about 2,000 first time students and 2,000 transfers; the college also served about 1,500 adults taking non-certificate, non-degree courses. Its ethnic composition was about one-third white, 24 percent Asian, 23 percent Hispanic, and 4 percent African American. The college serves a larger population of international students than most two-year institutions; eight percent of its undergraduate students in 2012 were from foreign countries. As expected, the majority of students graduating received a degree in liberal arts and sciences, but about 30 percent received certificates or degrees in health professions and related programs.

Interviews with early administrators revealed that they had worked hard in the 1980s to integrate the two types of faculties—the liberal arts and the vocational track. In cooperation with the state, steps were taken to increase the academic credentials of the vocation or workforce faculty—to enhance their status and their legitimacy. The early successes were in the health services area. Close ties were created with medical groups in the community to craft training and retraining programs for various types of allied health training programs. The demand for these graduates was steady even in times of economic downturn. In contrast, connections to the tech industries proved to be much more volatile. Administrators explained that these industries were focused more on high-end recruiting—engineers and executives—and less concerned about the technicians, about "what's happening on the floor." The college has had an engineering department for a number of years, but program reviews indicate that it has always been very small, at times employing less than one full-time equivalent instructor.

Foothill worked with a number of companies, including Hewlett-Packard, Tandem, and Cisco, to develop partnerships in which the companies provided equipment and Foothill provided training and other technical assistance, such as help with the writing of training manuals. However, administrator interviews indicated that many of these agreements broke down because the companies insisted that the training be provided only to their employees, whereas the colleges were required by law to have all of their courses open to the public. Over time, however, Foothill was able to encourage various categories of faculty to develop and cultivate advisory groups composed of faculty and appropriate industry groups, but also staff and students, to create programs serving practical as well as academic needs. A mission-based model of organization was developed to supersede the "silo" model—in which each department defends its autonomy. Today, Foothill sees itself as providing the "middle-skilled" workers to industry. As one administrator explained, "The companies will find for themselves the high-skill employees they require, but they also need the receptionists, desktop applications, technicians."

A second community college, *Evergreen Valley College* (EVC), is located in the eastern foothills of San Jose and is part of the San José Evergreen Community College District. EVC is a relatively new school, founded in 1975, more than 50 years after its sister institution, San Jose City College. An administrator explained that EVC spans two different communities: Silver Creek Valley, containing a cluster of millionaire homes and East San Jose, a low-income neighborhood. It currently serves about 9,000 students. The vast majority of students are either of Hispanic descent (40 percent) or Asian descent (39 percent). Tuition for in-state residents in 2014 was $1,300, and for out-of-state students,

$7,000. In that year it awarded 538 associate degrees, one-third of them in business fields.

Educational administrators reported that about 10 years ago the college shifted its emphasis from primarily academic to career readiness, including nursing, business administration, and automotive programs. For the last of those, the college has partnered with Honda and Tesla, which have production facilities in the region. The composition of the faculty has also shifted dramatically toward adjuncts; there are now 2.5 part-time instructors for every full-time faculty member. EVC's academic programs and completion rate are hampered by its high proportion of Hispanic students, many of whom are English as a second language (ESL) students. Over 60 percent of the college's students are currently employed. For these and other reasons, EVC has been less able to capitalize on its location in Silicon Valley than have San José State and Foothill. Compared to Foothill-De Anza Community College District, San José Evergreen serves an area with twice as many residents, has about twice the number of vice chancellors, but half the number of full-time students. Foothill-De Anza has about 237 acres at its disposal, while San José Evergreen has about 182 acres, which gives it significantly less room for growth.

Nonprofit *Menlo College* is located in Atherton, just north of Palo Alto. It became a college in 1927 when the private Menlo School for Boys expanded to include a junior college. During the 1930s there was discussion of having Menlo College serve as Stanford's lower division institution. While this plan was not adopted, Menlo College did create its School of Business Administration in 1949, just when Stanford had elected to drop its undergraduate business program (Curtis 1984).

Today, the business major is integral to Menlo's success; it describes itself as "Silicon Valley's Business School." The large majority of the college's degrees are offered in the area of business management, including specialized degrees in accounting, entrepreneurship, finance, marketing, and human resources. The School of Business Administration is served by an active industry advisory board, and in 2014 it received accreditation from the Association to Advance Collegiate Schools of Business (AACSB), a prestigious recognition obtained by only about 5 percent of business schools in the world. Total enrollment in 2014 was 794 students, of whom virtually all were full time. Published tuition is $37,000. Ethnic composition was 35 percent white, 21 percent Hispanic, and 7 percent each African American and Asian. Nonresident aliens made up 13 percent of the students. In 2014, Menlo awarded 160 BA degrees, more than

80 percent of them in business. Nearly 60 percent of Menlo faculty are adjunct. The college is proud of its one-semester exchange programs with universities in China, Japan, Spain, and Italy.

DeVry University, founded in 1931, is among the oldest and largest postsecondary educational institutions. From the outset, it has focused its efforts on career-oriented education, with an emphasis on technology, science, and business. During the period of our study, DeVry operated four Bay Area campuses: San José, Daly City, Fremont, and Oakland. We were unable to retrieve data specifically for these campuses, but based on IPEDS data (which are reported at the state level), DeVry served in 2014 a total of 7,932 students, 80 percent of whom were undergraduate, and 46 percent of whom were full time. The largest ethnic group enrolled was Hispanic (33 percent), followed by white (25 percent), Asian (14 percent), and African American (12 percent). Full-time tuition in 2014 was $17,000 plus $1,300 in fees. The great bulk of graduates in that year were in business and related fields (80 percent of bachelor's and virtually all of the master's degrees. The next largest major was computer and information services (21 percent) and engineering technologies (7 percent). DeVry has developed partnerships with such hi-tech companies as Google, Salesforce, Lockheed, and Logitech, as well as medical centers such as Varian Medical Systems.

Interviews with academic administrators revealed that less than one-third of DeVry's students enroll directly from high school; the majority are older adults. Classes are offered year-round, with many available during the evenings and weekends. All courses are taught each term so that students can always get the course they need. Curricular decisions are highly centralized and the college is highly sensitive to market signals, relying on both national and local advisory boards. Administrators report that when a market for training looks promising, courses can be developed and introduced within six to eight months. Local campuses work hard to build relationships with companies, holding career fairs to which industry representative are invited every eight weeks.

Advising and staffing structures at colleges like DeVry are different from community or state colleges. Much attention is devoted to developing very clear transfer agreements and to ensuring that students receive academic credit for past courses taken and work experience. Students are given clear advice about exactly what courses they need to take to complete a program. One interviewee explained, "We are not an exploratory university; it is rare for students to switch majors here." Each student is assigned a "success coach" who stays with her the whole time she is enrolled and provides guidance on issues as wide-ranging as transportation, financial planning, drug treatment, and mental health. An Academic Success Center is staffed by tutors available to assist with every class offered by the college.

Concluding Comment

Our approach is based on the premise that regional studies are an important and relatively neglected type of work. Still, we emphasize two caveats: first, regional connections are more fateful for some organizations than for others. Organizations—both companies and colleges—vary in their degree of embeddedness in their region. Second, as we have illustrated in this chapter, regions vary internally in their characteristics and composition. They differ demographically in ethnic makeup and economically both in terms of modal income and industrial composition. And, as we have emphasized throughout, colleges vary substantially in structure and mission.

The combination of these two sources of variation results in a collection of colleges that exhibit wide variation in their student populations, course offerings, and degree of successful adaption. Indeed, in some ways, the extent of the variety exhibited by colleges makes it difficult to arrive at valid generalizations regarding what the colleges are, how they are organized, and how they function. Nevertheless, in the next chapter, we attempt to characterize some of the common features detected in the ways colleges structurally adapt to their context and in the kinds of strategies they employ to relate effectively to their environment.

6

Structures and Strategies for Adaptation

W. RICHARD SCOTT, ETHAN RIS, JUDY C. LIANG, AND MANUELITO BIAG

As illustrated in the previous chapter, colleges have attempted to adapt in a variety of ways to both honor and preserve the traditions of higher education, such as degree programs, while adapting to the demands of a rapidly moving economy. Each college possesses unique structures and capabilities and confronts different kinds of challenges and opportunities, but, by looking across the full spectrum of schools, it is possible to detect some general patterns in the ways in which these programs work to cope with the demands confronted. We consider here some of the general strategies employed by colleges, including strategic leadership, structural differentiation, mechanisms linking colleges and companies, and modes of increasing flexibility. We also revisit some of the types of constraints that prevent their rapid adaptation. Since colleges vary in the types and strength of constraints they confront, we conclude by noting some of the advantages enjoyed by for-profit colleges, which exhibit greater agility than their public and nonprofit counterparts.

Administrative Leadership
Managing Complexity

We have called attention to the complex nature of the environments within which colleges operate, both within the sector of higher education and in the regional economies within which they reside. A long-standing truism of organization theory is that complex environments give rise to complex administrative systems in the organizations confronting them (Scott and Davis 2007: chap. 3). Administrators arose primarily to plan and coordinate the technical work carried out by employees, but as the wider systems in which their organization was involved became more complex, administrators were expected not only to manage internal processes but also to deal with external forces affecting the organization's functioning. For example, in colleges, special units were constructed to deal with the recruitment, enrollment, and placement of students; the provision and oversight of student services; the tending of alumni; negotiations with unions; coordinating with professional associations; and dealings with federal, state, and regional offices (see fig. 2.5). As these external relations have multiplied over time, the numbers of administrators have continued to grow, as has the proportion of administrators to other types of employees.

The numbers and types of external connections vary greatly by type of college. Also, their salience varies. Because all organizations must secure resources from their environments if they are to survive, the source of these resources or those that control them becomes of vital importance to organizations. Organizations become dependent on those who control their resources: unequal resource flows give rise to power-dependence relations (Pfeffer and Salancik 1978). For example, nonprofit colleges are more dependent on attracting students able to pay tuition and courting alumni and philanthropists who offer donations. Public colleges cultivate their relation with legislatures and public agencies that fund and oversee their programs. Research universities are dependent on federal agencies and, increasingly, on corporations that can support their inquiries and graduate training programs. And for-profits adjust their curriculum to meet changing market demands in order to satisfy their investors. In all of these cases, mission is closely attuned to money, and changes in the sources of revenue can affect the operational mission of the college (Weisbrod, Ballou, and Asch 2008).

Strategic Leadership

One hundred years ago, there was less controversy over what the "mission" of a college was. It was largely taken for granted. In the modal college, administrators primarily played a support role to faculty, and took it upon themselves, without much deliberation, to enact the traditional role of the college: to preserve, enlarge, and pass on the scientific and cultural heritage of our civilization. For many years, university officials were expected to safeguard the legitimacy of the college by adhering to the prevailing organizational templates and practices: isomorphism was the key to legitimacy. Over time, however, as competitive processes have become ascendant, administrators have come to play a more dominant role in crafting and marketing an image or identifiable profile for their college—a "brand" that highlights their distinc-

tive features and programs. Rather than embracing a general institutional account of their role as a university, they increasingly feel compelled to develop
a customized organizational mission that sets them apart from their competitors (Hasse and Krücken 2013). The rise of rankings and ratings by national
and international media has fueled this invidious comparative process. Like
other competitive entities, universities are likely to hire professional consul-
tants and to contract out many of their services to private providers. Under
heightened competitive conditions, many college leaders have decided to pursue what Suchman refers to as "pragmatic" rather than "moral" legitimacy.
Moral legitimacy refers to actions that are regarded as the right thing to do in
that they are believed to effectively promote societal welfare. Pragmatic legitimacy, in contrast, takes into account the "self-interested calculations of an
organization's most immediate audiences" (Suchman 1995: 578–79; see also
Gumport 2000).

The administrative component of colleges has become larger but more centralized, while the number and proportion of tenured faculty members has
declined. Data compiled by Weisbrod and colleagues (Weisbrod, Ballou, and
Asch 2008: 202) show that since 1993, the earliest year for which tenure-track
data are available, there was approximately a 10 percent drop in the use of
permanent faculty in both public and nonprofit schools. (For comparable data
regarding changes in the Bay area, see table 3.2.) During this period, public
schools' tenured professors dropped from 59 to 49 percent. These figures indicate not only increasing cost-consciousness on the part of colleges over time
but also the likelihood of a diminishing role for faculty in setting the priorities
of their college or university. Nontenure or adjunct faculty have no voice in
college decision making. Moreover, as noted in chapter three and amplified
below, they afford administrators flexibility in creating and reconfiguring program offerings to conform to changes in mission or demand.

More generally, the caretaker or support role formerly assumed by administrators has been supplanted by an entrepreneurial conception that stresses
that the central role of the chief executives is to craft organization strategy—
not simply to respond but to anticipate new demands. Virtually all of the recent
emphasis has focused on the importance of meeting economic, not scholarly
or more broadly societal, demands. There has been an explosion of writings
noting this dramatic shift in the view of what colleges are and how they
are expected to behave. Some of this literature is descriptive, detailing recent
attempts—both more and less successful—to craft a dynamic engine of

economic growth. Other contributions are either hortatory, exhorting executives to take command, or critical, condemning the incursion of market forces
into the temple of education (e.g., Berman and Paradeise 2016; Clark 1998;
Etzkowitz 2003; Fayolle and Redford 2014; Foss and Gibson, eds. 2015; Lowen
1997; Slaughter and Leslie 1997; Washburn 2005). Not unexpectedly, the great
bulk of this work singles out the research university as the primary subject.
This type of organization looms large as one of the main sources of new knowledge which can become the basis for commercial science and technology; the
research university also can be a source of top business and engineering talent,
which can transform ideas to models to prototypes to products, providing the
groundwork for new companies or streams of revenue. And an important fact
is that universities stand to benefit directly from these products in the form of
increased research support, patents and licensing, shared profits, and gifts
from business collaborators and alumni.

But, as we have emphasized throughout, higher education does not consist
only of research universities. However, these types of organizations often
provide the prototype for other educational institutions, so that we observe
market forces permeating and affecting the development of the full range of
colleges. Not only university presidents but also presidents of four-year and
community colleges and nonprofit liberal arts colleges have been obliged to
adopt new leadership styles. For example, we noted in chapter five that San
José State University has adopted as their official aphorism "Fueling Silicon
Valley," while the nonprofit institution Menlo College refers to itself as "Silicon
Valley's Business School." We talked to one administrator in a community
college who recalled that when she became president she was confronted by
individual departments and programs, each with its own procedures and requirements. This administrator took steps to disrupt this "silo-model" with
general "mission" models, each of which cuts across traditional academic lines.
Programs were crafted and resources devoted to serve these broader missions,
and an attempt was made to create rough parity between the missions. For
example, career education was treated as of equal importance to the transfer
(liberal arts) mission. A high-level administrator from Foothill College describes efforts at both the state and college levels to give parity to workforce
development programs. The minimum qualification for faculty in these programs was raised: an MA degree was required and, internally, efforts were
undertaken to increase their status and legitimacy.

Administrative leadership has many facets. Strategic vision is both of prime importance and relatively rare. It involves identifying opportunities as well as assessing issues or problems that require attention. It requires awareness of the organization's competences and limitations, and the ability to mobilize people and resources to respond appropriately. We found evidence of such leadership in some of our sample colleges. For example, an informant told us that although computers were central to the South Bay that the important economic players in the East Bay were in biotech and logistics, serving transportation and port facilities. Another pointed to important emerging areas in which colleges needed to be involved, such as nanotechnologies and cybersecurity. An administrator reported,

> I always look for my faculty champions—those with expertise and passion—who will drive us forward into new areas. I assembled a STEM advisory board, an interdepartmental group, charged with helping us design curriculum that would really be relevant and would help our students succeed. Nanotech was one area they recommended. That group is also going to be responsible for a quality assurance skills program that will start this fall. Part of what we need to do as a community college, since we can't move in and out of all the specialized training, we need to make our students life-long learners with resilience and persistence and some grounding in the foundational knowledge.

And a second administrator remarked, "There are top-down initiatives that come from the Office of the President or the Provost that are cross-disciplinary. As an example, a new faculty learning community that is emerging is clustered around cybersecurity and big data. That crosses disciplines. We are bringing in a faculty learning community cohort; this way it's not siloed. Another example: we are introducing gaming as a major and develop its use as a learning tool. So what's happening in the Valley does affect many programs here." More generally, another administrator described his strategic approach:

> It's about understanding what the demands of the community are and building your curriculum around them, but also being adaptable to address things in the future. Adaptability: that's the key thing. Every year I talk to my faculty and staff on opening day and tell them: "What I ask of you is to be adaptable. Don't commit to death to the things you know: be comfortable dealing with ambiguity. Recognize that taking measured risks pays off. And in reality they aren't

even measured risks. They are things that are really designed to address the broader needs of our students.

Identifying capabilities within the organization that can be exploited is also an important part of strategic leadership. Another community college administrator explained how student abilities could be leveraged:

> Seventy percent of our students speak two languages at a minimum: 60 percent of the auto technicians speak two languages; 70 percent of the nurses. If you are running an auto repair shop in the community and you've got a mechanic that speaks Mandarin or Vietnamese or Spanish, you've just struck gold! Is that ever in the job application? Is that ever part of the application process? The regional director of [a large automobile company] responsible for hiring didn't recognize this: I had a conversation with them years ago and you could see the light bulb turn on. That's totally cool! If you have somebody on the floor that can talk to clients, the old Vietnamese lady, it's good business practice.

Administrative leaders with strategic vision are vital to the health of organizations, but such individuals are scarce in broad-access colleges. One administration described his view of the situation:

> The single biggest problem in the community college system is the lack of really good administrators. We grow really good faculty and classified staff but administrators are hard to grow because the demands on the job have increased so much. I remember talking to a guy who was a dean back in the early 1990s. He was acting dean for a while when his dean was away for a year and he told me that the changes that had occurred were unbelievable. You have to know everything. If you are a faculty member, you'd have to be insane to move into administration. Another consideration: when people are considering retiring in Silicon Valley, if you're good at what you do, there are a lot of good positions out there [in industry] for more money than you can get by staying in the college system.

Differentiation and Loose Coupling

Another time-honored method of dealing with an increasingly complex environment is to design a differentiated organizational structure. Differentiation occurs both around functional differences—for example, recruitment, admissions, teaching, counseling, placement, administration—and around product differences—for example, liberal arts, workplace or vocational training, professional preparation, remedial programs, adult education. The latter, because they involve less interdependence among employees and, hence, require less coordination of activities, are more likely to be autonomous: they are loosely coupled to one another and from the larger organization. Functional departments are replaced by, or are incorporated into, product divisions.

The most complex of the college populations—research universities and comprehensive colleges—display highly complex organizational structures. Academic departments enjoy considerable autonomy in their construction of curricula and requirements (although, as described in chapter two, they operate within the frameworks set by and under the watchful eye of external professional associations); and instructional programs for workforce training, adult education, and professional training take place in separate programs or schools that may have loose connections to the universities within which they operate. This structural arrangement provides the flexibility necessary to allow the college to simultaneously pursue diverse and sometimes contradictory objectives. As we have seen, Stanford's and San José State's Schools of Engineering are able to mount unique and innovative degree programs and admissions criteria and to independently seek financial support from companies and foundations (see case examples 3.A and 5.C).

As broad-access colleges have shifted more of their attention and resources to vocational and professional preparation, they have been obliged to strengthen these programs and to increase their status. As a community college administrator reported, "Lots of community colleges saw themselves first as transfer institutions and then as workforce developers. The biggest question was the status of workforce development programs on the campus and the status of faculty, and what commitment do colleges make to work with industries versus universities." In association with the passage of the AB1725 bill in 1988 in the California legislature to "elevate" vocational education to be of equal importance in the mission of the community college system (Livingston 1998), he continued, "we worked toward an equalization of the workforce programs and the transfer programs so that we don't see that two-tier system as much as before. . . . Look, these colleges are all enrollment driven: you need to be savvy; you had to recognize the transfer potential populations and to recognize the workforce groups, which had to be brought up academically to a level comparable to others." Although the 1988 legislation was intended to ensure that vocational education was of equal importance to academic education, the same bill required that 75 percent of credit courses be taught by full-time faculty, most of whom were not in vocational programs. This requirement was seldom met, but it did put pressure on community colleges to raise the minimum qualifications for all faculty members.

In some cases, faculty commented on changes over time in the orientation and leadership style of incoming university presidents. A new president of one of the state universities was reported to have ramped up the "decision-making process to lightning speed. . . . He's driving the provost in this way as well. In the year-and-a-half since he's been here, there's a lot of new pressure on faculty to change and adapt. We have a much more entrepreneurial culture since he took the helm, with lots of incentivizing for faculty to develop and build capacity in these [STEM] areas."

To oversee these differentiated systems, college leaders have been compelled to master "organizational ambidexterity": a capacity to guide programs serving diverse missions. Many college programs, such as the humanities and social sciences and many of the professional school courses, are relatively settled: curricula are set and faculty are trained to carry them out. These programs can be "exploited" by the college in the sense that they meet a steady demand and generate predictable revenues. Other programs are experimental and risky. Those on the cutting edge of science or involved in developing online offerings represent opportunities for colleges to "explore" new arenas of study or modes of instruction. Leaders of organizations such as colleges operating in areas experiencing rapid change must attempt to balance a portfolio of activities combining exploitation and exploration (Tushman and O'Reilly 2011).

Change by Selection

Organizational ecologists point out that change in organizations occurs through two fundamental processes: by selection and by adaptation. In selection, change occurs either as some organizational forms cease to exist or as new organizational forms are created. In chapter two, we identified six major types of college. Of these six forms, all have continued to exist during the period of our study: no forms have become extinct during that time. However, two of the forms—community colleges and for-profit colleges—have experienced relatively rapid increases in number of organizations and size of enrollments.

Community colleges have been transformed from marginal players to mainstream providers of higher education services. Dating from the early years of the twentieth century, community colleges began to expand in the 1960s and have continued to grow to the point where they currently enroll more than half of the students in higher education. This educational form is notable in attempting to combine within a single form attention to both the liberal and the practical arts in its offerings (Brint and Karabel 1989), so in that sense, it embodies the changing logics of the field.

Like community colleges, for-profit colleges have long existed in the periphery of postsecondary education, providing evening courses for working-class occupations such as beauticians and metal workers. But, following the breakout success of Phoenix University, founded during the 1970s by the institutional entrepreneur John Sperling (see case example 2.B), this form "blossomed into a full-service industry that is expanding in numerous directions" (Tierney and Hentschke 2007: 16). Largely unresponsive to long-existing support and control systems, such as professional or administrative educational associations, unencumbered by traditional notions of faculty autonomy or involvement in governance functions, for-profit forms have wholeheartedly adopted the corporate model embracing centralized managerial control in the service of shareholder interests. The form differs in fundamental ways from traditional modes of academic organization. This population has grown rapidly if fitfully during recent decades with periods of rapid growth punctuated by scandal and instances of large-scale corporate failure.

Selection forces operate in response to large-scale changes in supporting environments. Community colleges began to multiply rapidly in the later 1960s in the United States because of changes in public policy that unleashed public resources to rapidly expand access to higher education for large numbers of previously underserved types of students. The lion's share of resources deployed by states was expended in the construction of community colleges. By contrast, for-profit colleges began their rapid ascent during the 1970s because of the growing acceptance of the neoliberal ideology celebrating the role of market processes in meeting social needs. Corporate managers employed advanced marketing techniques to survey potential markets and rapidly developed courses responsive to local demand and made readily available at times and in places convenient for potential markets.

Adaptation Strategies

Unlike selection processes, adaptation involves changes that existing individual organizations make in order to better suit the needs of the environment. Colleges of varying types employ a wide range of strategies to adapt to the environments

in which they operate. Among the most important of these are bridging or spanning strategies and those that increase the flexibility of the institutions.

Building Bridges

When new players arise controlling valuable resources, organizations confront the choice to either attempt to "buffer" (protect) themselves from their
effects or to create "bridges" or spanning units to co-opt and/or collaborate
with them (Thompson 1967; Scott and Davis 2007: chap. 8). Bridges create
frameworks for nurturing and managing resource flows between organizations.
As public colleges have confronted reductions in funding from the state, they
have been obliged to seek new sources of revenue. Up to this point, the primary
sources of new revenue have been increased fees and tuition and support from
industrial partners. Both sources increase the pressure on colleges to be more
responsive to the needs of their "customers"—both students, who increasingly
expect their colleges to provide them with employable skills, and businesses,
which want qualified workers. Although we recognize that there are multiple
types of skills and employers, including those associated with nonprofit organizations and public agencies, our research concentrates on those most relevant
to serving the needs of firms in hi-tech industries. What are the types of bridges
colleges have constructed to connect to these firms?

We observe changes over time in the strategies employed by colleges to collaborate with industries. During the 1970s and early 1980s, many colleges received *equipment from companies* and, in return, provided training to some of
their workers. Some of these programs worked well. One community college
was provided with computer reservation (Saber) systems. Saber donated the
hardware and software to support the creation of a program in travel careers
with connections to travel agencies that was quite successful. Another successful example of college-company partnership was the program developed by
Foothill College and Tandem Computers and then Cisco. Foothill persuaded
Tandem to donate computers in return for which they provided training to
students and helped Tandem rewrite its training manuals. Foothill taught
courses to Tandem employees as well as to potential employees. Tandem
officials came to the college to hand out certificates to successful graduates. "This connected folks together; connect students to employers to see
whether there was a match," an administrator recalled.

As a general strategy, however, this approach proved unsatisfactory because
the technology changed so rapidly that dedicated classrooms quickly became

obsolescent. Another barrier to these ventures quickly surfaced for public colleges. Companies supplying the equipment were concerned about security and proprietary issues and wanted the use of these classrooms to be restricted to their employees, whereas public schools were obliged to make their classes open to all students. Early connections also relied heavily on *contract courses* in which colleges would contract with individual companies to provide specified types of training. Because of the open enrollment requirement for public colleges, these arrangements have been more commonly employed by nonprofit and for-profit schools. However, public colleges often participate in off-site contract courses or even entire degree programs by carrying them out through their extension divisions (see case example 3.A). For example, San José State University offers a variety of off-site MA programs taught within companies for their employees.

A more widely used bridging strategy has been the development of *internship* and *apprenticeship* positions with colleges. The former is more common for semi-professional and technical training such as nursing and dental hygiene, the latter for manual, "dirty-fingernails" positions (as one respondent described it), for instance, plumbing and automobile mechanics. Colleges such as Evergreen have shifted more of their resources to training students for nursing and automotive training. However, there are exceptions. Program reviews show that Skyline College's biology department has operated a robust internship program since 1991 as well as a seminar series featuring industry speakers that dates to 1975. Students are placed in these positions to give them hands-on training and experience for which they receive course credit. Some of these positions are for pay, others not. They often provide pathways to employment in the same or a related job. However, such positions appear to be less common in many of the hi-tech firms, in part because these companies are more inclined to hire temporary workers who, if they perform well, can be shifted to permanent hires.

As we will see, *adjuncts* fulfill a number of functions for colleges and companies, but one of them is certainly bridging. Adjunct faculty are often either current or recent employees who bring with them current knowledge of the contemporary workplace and practices. They can share information with their students about state-of-the-art software and other technical information and know-how. In addition, adjuncts can spot promising students and funnel them to opportunities within the industry. A representative from the San Mateo Community College District reported that many schools in the San Mateo and South Bay regions rely heavily on technology firms such as Google, Yahoo, and Hewlett-Packard for adjunct faculty. This was echoed in program reviews for many computer science and engineering departments, which have actively recruited practitioners from these and similar companies to teach students industry-specific skills. Respondents noted that adjuncts—unlike tenured faculty who may have been hired years ago and failed to keep up with industry changes—have applied knowledge about the types of competencies now current that will position students competitively in the job market. Moreover, practitioners have connections with professional communities that help students land jobs after completing their training. Still, many informants noted in program reviews that they found it difficult to compete with industry in terms of salary for adjunct instructors, given budget constraints.

In the early years, some of the larger companies in the Bay Area such as Hewlett-Packard appointed company managers in various departments to serve as *points of contact* for specific colleges. These managers were typically graduates of the programs they were expected to coordinate and so could make use of their previous contacts and experience to cultivate and grow their network connections between company and college. Many of the managers were volunteers who were motivated to act as loyal alumni, providing qualified students for their alma mater and, at the same time, advancing the qualifications of their company's employees. Our sense is that these programs were effective for a short time but became quickly outmoded.

Without question the most widespread and, probably, most effective mechanism for bridging companies and colleges are *advisory boards*, in which industry representative are appointed to regularly meet with college administrators and faculty to provide advice and assistance on courses and programs. Most of these boards involve collaborations within the region, but some for-profits also convene national business advisory boards because their curriculum planning is centralized at this level.

In the 1970s and early 1980s, advisory boards were rare. They were typically the result of actions taken by one faculty member or department chair looking to develop stronger ties with industry and seeking guidance on current needs and trends. As these proved to be productive, particularly for colleges, administrators including deans and faculty began at first to urge and then to require their industry-related programs to develop such connections. Indeed, advisory boards are required as a condition of certification by many of the specialized professional school accreditation bodies. Hence, these boards are now widespread, particularly in the STEM branches of research universities and comprehensive colleges, and they function in numerous ways. Many advisory boards include in their numbers alumni of their college. These representatives have a vested interest in helping their alma mater improve, and many of them also serve as placement conduits for students. Some colleges invite their board members and representatives from other companies to sit in on and react to student presentations. Others link these types of functions directly to work done within companies. For example, in one CSU college, advisory board members would attend meetings where students did "senior design presentations in which a group of students go to some company and solve problems for them. . . . Board members can then tell us what are the weaknesses and the strengths of this work and what they want them to know, including communication skills. And then we decide in those meetings what to change in the curriculum and software that they are using that we can add to our training."

In some cases, direct monetary support from companies is sought, but the most common coin of exchange is information. Specifically, colleges turn to their advisory boards for information regarding their current and expected future needs, including particular skills and types of knowledge as well as new types of curricula and course offerings.

In the early decades of attempting to respond to the needs of Silicon Valley, colleges often attempted to introduce entirely new courses, often linked to the introduction of new types of hardware and software technologies. However, as noted above, particularly for public colleges, by the time a new course could be mounted, the need for it had passed. The compromise adopted by some of the more responsive and nimble colleges in our sample was to modify existing courses rather than adding new ones. Current courses could be amended to include new curricular nodules and/or new laboratory components and exercises. These types of changes had the advantage of not requiring approval by central administrators, bypassing this time-consuming approval process.

In addition to creating or serving on a specific college advisory board, some college administrators, including presidents, and many company managers serve on one or more regional research and/or policy board, such as the Bay Area Council Economic Institute or Joint Venture Silicon Valley (see chapter three). Membership on these advisory bodies provides an opportunity for mutual learning and problem solving among all participants.

Another historically important form of bridging for colleges has been through the activities of their *placement units*. In the past, placement departments within colleges have served as central units connecting students with prospective employers. In the early years, these units were well staffed, providing counseling to students about job opportunities, resume preparation, and training for job interviews. Frequent job fairs were hosted and space provided for company representatives to meet and interview job applicants. The School of Engineering of San José State continues to operate an active career center. A respondent reported that "We graduate 600–700 engineers per year and most stay in the Bay Area, so the University and the career center in particular knows who is looking and who is hiring. We're constantly watching." However, budget cuts in recent decades have eliminated much of this capacity in many public colleges. And, as described in chapter three, online placement services such as LinkedIn have increasingly become the go-to mechanism for connecting employers and workers (see case example 3.B).

Since many of the colleges in the Bay Area region are attempting to develop linkages with companies, competition among them has increased, and some are necessarily more successful in developing partnerships than others. More obviously, some departments and programs are better suited to collaborate with industry than others. Programs of study such as digital media are better suited than anthropology or classical studies to meaningfully connect to industrial companies. Stanford University is attempting to develop collaborative connections to allow humanities and arts programs to partner with computer science and engineering, but few colleges have the resources to facilitate such programs. Capacity to meaningfully partner with hi-tech firms varies greatly by region, college, program, or department.

Colleges can also form productive linkages to one another. A strong example is provided by allied health programs developed at Foothill community college which contracts with the Stanford medical school to collaborate in developing program content. While paying Foothill tuition, students in one such program take basic educational courses at Foothill and then take physician assistant courses at Stanford and are taught by Stanford medical school staff. A Foothill administrator noted that such cooperative arrangements in the health care field were much easier to develop and maintain than those with tech companies because health care is less subject to rapid technical obsolescence and economic volatility.

In addition to the ties or bridges to specific companies or colleges, numerous colleges have participated in *organizational networks* many of which operate at the regional level. Several of these involve Linked Learning, the new California Career Pathways Trust, which supports efforts to make more seamless transitions from K–12 to postsecondary institutions and jobs. Other networks involve the adult regional planning effort created by AB86, which provides funds to regional consortia to create plans to provide adults with basic skills and technical education programs with high employment potential. Other programs include those developed within the Division of Workforce and Economic Development within the Community College Chancellor's Office to identify emergent industry sectors, such as health care, information technology, and tourism, in order to assign "Deputy Sector Navigators" to work with colleges so that they can better prepare students with the skills needed in the twenty-first century.

Increasing Flexibility

We have repeatedly emphasized the rapid pace of change in the Bay Area economy. As the economy shifts to accommodate changing demands, companies and workers must attempt to quickly transform technologies and work practices. For reasons including regulatory controls, professional norms, academic cultural traditions, funding constraints, and inertia, many colleges, particularly public colleges, are not able to easily accommodate to demands to provide new and different types of knowledge and training. In our macro studies of academic programs and degrees, we have observed changes over time on the part of most types of colleges (see chapter three), but these changes are measured in years and decades, not in days or months. In our up-close studies of selected colleges, we have also obtained information on some of the many ways in which colleges are attempting to adapt to market pressures.

Diverse Programs and Missions

As we have emphasized throughout, the varying populations of colleges serve multiple diverse missions or goals, ranging from remedial education to research and professional training. We regard the wide variety of missions being served to constitute one of the great strengths of the US system of higher education. It means, for example, that a much greater number of students varying widely in income, previous education, ethnicity, and age can be exposed at some level to the college experience. These diverse missions are distributed throughout all the types of colleges, but even within a given type of college varying programs and training opportunities are available to students. For example, all community colleges offer traditional liberal arts transfer programs but also provide CTE—career and technical education. Such courses range from auto mechanics to computer programming and paramedical training. In addition, many offer GDE programs to part-time students seeking to earn high school diplomas. Community colleges provide ESL (English as a second languages) courses, affording non-English speakers an opportunity to move into the mainstream. And most community colleges are compelled to devote time and resources to remedial education, offering courses that are below college level to allow students to build basic reading, writing, and computational skills. While this activity represents a failure of the K–12 systems, such programs provide many students with a second chance to enter genuine college courses. State universities (comprehensive colleges) offer an even broader range of courses to their diverse student bodies.

Diverse Emblems of Achievement

For many years, the gold standard of academic achievement has been the college degree. The difference between having taken some college courses or being experienced and well read on the one hand versus owning a college degree is substantial, as measured by being eligible versus ineligible for a wide variety of occupations and differences in earning power (see Hout 2012; Meyer 1977). As described in chapter two, an elaborate apparatus of accreditation exists to determine both what kinds of educational systems can offer degrees and what programs of study need to be completed by individual students to qualify for this emblem of achievement.

More recently, spurred in part by programs initiated at marginal postsecondary institutions such as trade schools, mainstream colleges are offering a wider range of credentials, including certificates and badges that recognize abilities in specialized fields of competence (see chapter three and case example 3.C). This move reflects in part the desire of education systems to find ways to recognize the achievements of the diverse types of students being served in programs offered. Faced with increasing accountability pressures, colleges have embraced the development of measures other than degree completion. In their discussions with us, a collection of community college leaders described their frustration with the limited range of official measures employed to evaluate college performance. They pointed out that for many of the types of students enrolled, measures of success should be expanded to include numbers of GED awarded, successful completion of remedial courses, increased proficiency in English, elevated educational aspirations, and the achievement of one or another type of specialized certificate.

Many types of employers indicate that they are more interested in hiring workers who have specific types of competence rather than general knowledge. Badges and certificates are helpful emblems for both these employers and workers. Workers can report such achievements on their online applications and employers are able to select for them in their recruitment. And, as described in chapter three, a variety of approaches to developing broader frameworks are being proposed to standardize the types of certificates awarded and to validate their claims. But it remains true that for many categories of positions, employers seek applicants with college degrees, as evidence of general knowledge, analytic skills, and social capital.

Enrollment and Scheduling

For most degree-granting colleges, the academic calendar has remained largely unchanged. Applicants apply months in advance for acceptance; courses are scheduled across several months, broken up into either semesters or quarters; and, to receive credit, students must remain in class through end-of-term exams. Classes begin and end at fixed times, convening on a fixed number of days throughout the term. It once was true that the modal college student attended full time, but this has now changed, and gradually other changes are being introduced to render colleges flexible and convenient for students.

Because of the volatility of the Valley, numerous colleges, including both community colleges and state universities, have initiated *intermittent enrollment policies* that allow students to leave and then resume their educational programs without having to reapply. Rather than treating these students as dropouts or in an ad hoc manner, colleges such as San José State have instituted more flexible enrollment policies to accommodate them. Such programs can also adjust to changes in student priorities. Informants told us that "during the dot-com boom of the late 1990s, students would come just for the specific skills they needed since jobs were plentiful. Then, when the Valley crashed, they came back to school, but this time seeking a degree." Describing the situation, another informant reported: "We're like a bus station. There's a time table and a ton of transferring and coming and going. It's such a different culture than Stanford where 90 percent of the students who come in as freshmen graduate in four years." And another: "The students that come from the community colleges and transfer here—some 40 to 50 percent transfer—have on average been to two to four community colleges already." And, an increasing number of colleges permit part-time students. Colleges have also attempted to streamline and more liberally evaluate and accept transfer credits. Many colleges have developed transfer agreements with other colleges, but these arrangements remain too limited.

Another strategy for making college courses accessible is to extend the *hours* and the *places* where they are available. Numerous community colleges, state universities, and nonprofit colleges have developed branch campuses, often closer to urban centers, and they have also introduced evening courses for students who work or have other commitments during the day. Also, as previously described, some research universities have developed extension programs that attempt to provide students with maximum convenience in terms of location and timing (see case example 3.A).

Maximum convenience to students is offered by *online courses.* Such programs are slowly growing in the mainstream public and nonprofit colleges, as described in chapter five, but remain a relatively small component of their curricula (see also case example 4.C). As discussed below, they have been more widely adopted by for-profit colleges.

Adjuncts

We have discussed the increasing use of adjuncts throughout this book, but emphasize here their contribution to college flexibility. As college costs have increased and revenue becomes harder to come by, colleges have reduced the proportion of tenured faculty members, turning to the use of short-term, part-time instructors. A substantial number of these instructors teach in liberal arts programs, many commuting among several schools, teaching a class or two at each. They not only lack tenure but also benefits and, in most cases, even an office. Their pay is low (Bradley 2004; Washburn 2005). For colleges, however, employing this type of faculty allows them to rapidly expand or contract their offerings to suit changing enrollment and curricular demands.

Of more interest for our research purposes are the part-time instructors who come to the college from industry backgrounds. Many of these are recent or current employees, who are interested in teaching evenings after completing work. As noted, they form important bridges between colleges and companies, bringing knowledge of current needs and practices into the college, helping to link students with jobs and companies with promising students. And, as companies, technologies, and work processes change, new company employees can be hired to replace earlier ones. Across our sub-regions the need for and availability of adjuncts vary considerably, but all of the colleges we examined made at least some use of this valuable resource.

Administrators and faculty at several colleges pointed out the costs entailed in utilizing adjunct faculty. Each adjunct needs to be identified, recruited, evaluated, and provided with some orientation and guidance regarding college procedures. Effort is required to plan programs in advance, to determine what kinds of adjuncts are needed and what kinds of laboratory or technical assistance they will need. Someone must exercise supervision and attempt to maintain quality standards. Someone must pick up most of the advising and guidance responsibilities for such faculty, since they are only on campus for a short period surrounding their classes. Administrators assume some of this burden, but much of it falls on the dwindling numbers of tenured faculty members. At the same time their numbers are declining, the administrative duties of tenured faculty are rising, affording them less time to devote to their own teaching and scholarship.

Online Courses

The ultimate in always-available-on-demand education is represented by the development of online courses, which reached its apogee in the first years of this century with the launching of OpenCourseWare by MIT, quickly followed by the emergence of several major providers of MOOCs (massive open online courses; see case example 4.C). These courses are accessible to all students with access to the internet, and a substantial number of them are free. Because of their low costs and convenience to users, many predicted that this new product and delivery system would represent the end of traditional educational providers. That this has not happened can be explained by several factors.

- The courses developed to this point make heavy demands on their users, who have to already be literate, disciplined, and capable of sustained self-motivation over the duration of the course. Most of those who have enrolled in these courses fail to complete the course, and those who have been more successful typically have previous college experience if not degrees. Large numbers of students currently being served by community colleges lack sufficient educational preparation to benefit from online courses.
- Most of the courses being offered are not linked to or backed up by a certification system that verifies that the class has been successfully completed. As described in chapter three, these systems are still in the early stages of development.

- The modal college student does not view an online education as an adequate alternative to a "real" college experience, which features not simply cognitive development but interpersonal interaction, emotional development, and participation in a learning community that enhances both intellectual and social capital.

We have noted in chapter five some attempts by community and comprehensive colleges to make use of online programs to create more blended courses. To date, such approaches, while promising, remain limited. In contrast, as described in chapter four and below in this chapter, for-profit colleges have been early adopters of this new technology and have made much more extensive use of online courses than their public counterparts.

Barriers to Adaptation for Public Institutions

Throughout our discussion, we have emphasized the important differences that exist among types of colleges. Public colleges have historically been and remain the main provider of educational services in this country, and they, more than other types, experience greater oversight and stricter controls. Nonprofits and, especially, for-profits operate under fewer constraints. We briefly review three types of controls on colleges.

Administrative and Political Controls

We described in chapter four the types of administrative and political controls in place in California. All public colleges comply with legislative mandates and administrative policies, which vary considerably by type of college. All receive substantial revenues from state sources and are subject to the vagaries of state economic conditions, changing legislative priorities, and shifts in the broader political climate. It is not surprising, for example, that the decline in state appropriations for higher education coincided with the antiwar student protests on college campuses during the late 1960s. More generally, college authorities confronting unpredictable budgets and changing priorities have difficulty making rational decisions. In addition, state administrative oversight is elaborate and its requirements are often outdated and cumbersome. For example, new course proposals require approval from state authorities, which occasions delay. As one community college informant told us,

> There's an entrepreneurial spirit within the college and we've gotten into some trouble getting creative within the state system and its strictures. This is illus-

trated when you look at what it takes for us to get a course approved. We have nineteen new computer science classes to offer this coming fall. They were ready a year ago, but the approval from all the state agencies and the CCCO (Community College Chancellor's Office) took a year. I'm sure some aspects of these courses are already outdated . . . so there's naturally a mismatch, especially when it comes to curriculum approval.

Public and many nonprofit colleges also employ a relatively distributed decision-making structure, authority being shared not only between colleges and their political/administrative systems but also within colleges, between administrators and faculty. Although the extent of faculty participation in university decision making varies substantially among colleges, it has been in decline for several decades. Still, a considerable amount of autonomy and discretion continues to reside at the school and department rather than the college level. For all its advantages, distributed authority is the enemy of rapid organizational adaptation.

Professional and Cultural Controls

The most important component of a college remains its faculty. And in spite of all the changes that have occurred, faculty members are professionals who are highly attuned to the requirements of their academic disciplines. As noted in chapter two, for most faculty members, academic identity trumps loyalty to a specific college. While academic disciplines are undergoing change, they continue to require long periods of socialization for their members, adherence to common standards, and an insistence on the importance of collegial controls. Adjunct faculty, in particular, those coming from industry, are much less supported or constrained by these controls, but they are obliged to function in systems that, for the most part, continue to adhere to them.

More broadly, cultural and cognitive systems play a continuing role in reining in rapid change within colleges. Not only faculty but also students and their parents share a strong common sense of what a college is for, the importance of education, and, especially, the value of a college degree. It continues to be true that even in community colleges, which are under strong pressure to push vocational programs, most of the tenure-track faculty and a substantial proportion of students seek to participate in a liberal arts transfer-degree program rather than a terminal vocational alternative. The major component of the faculty and, more important, the bulk of entering students continue to value and seek to acquire the long-term benefits of a liberal arts education. Important constituencies seek to undercut the forces of adaptation and change in order to protect an earlier set of values.

External Constituencies

In addition to the professional and academic associations surrounding colleges and their participants, other influential constituencies include union organizations and the alumni. Both of these forces are primarily conservative in their effect, engaged in resisting change. Unions, involving faculty, staff, and graduate students, work to preserve job security and benefits for their members. Alumni to a large extent are loyal to their alma mater and resist most types of change that threaten to alter the college they fondly remember. More important, they seek to ensure that their own children will enjoy the opportunity to experience the same social involvement and, perhaps, the intellectual excitement they fondly remember.

In sum, constraints on change vary. Some impose unnecessary obstacles to constructive improvements. Others, however, seek to preserve precarious values that have historically been served and which, in the eyes of many, remain valuable.

Openings for For-Profits

As we have documented, traditional public and nonprofit colleges are not constituted to rapidly adapt to changing contexts, whereas this is one of the major strengths of for-profit colleges. While for-profits enjoy important adaptive advantages, they have also exhibited the limitations of their mode of operation, shortcomings that have prevented their rapid increase in market share in higher education. From the 1970s well into the 1990s, the numbers and enrollments of for-profit colleges rose steadily, but in the early decades of this century, these increases have been slowed and, in many cases, reversed. This sector of higher education exhibits high volatility up to the present. But a word of caution. Our comments refer almost exclusively to that subset of for-profits that, perhaps in addition to other types of programs, offers academic degrees. These are the programs that are captured by the official educational data systems. For-profits offering exclusively vocational training and certificates operate under the radar screen, and an investigation into them would require a different design and data-gathering strategy than that pursued in this study.

Our interviews with sample colleges confirm that for-profit institutions continue to be key players in the field of higher education. Over time, they have offered a wide spectrum of two- and four-year programs (e.g., criminal justice, education, retail hospitality). Following changing market demands, they alternate between an emphasis on employer-guided vocational training and academic degree programs. As described in chapter five, colleges such as the University of Phoenix and DeVry offer more focused programs and provide a range of supports, such as "success coaches," financial advisors, counselors, and staff to support online work and, more generally, student learning. These distinctive advantages, however, are offset by the heavy emphasis placed by many of these systems on marketing. To succeed, colleges must attract new students (customers). Too many for-profits have employed sales techniques that overstate successful outcomes, including time-to-graduation or employment prospects (see case example 5.A). Some have resorted to outright fraud. For example, California's state attorney general levied a $30 million fine against Heald College (with four campuses in the Bay Area), a branch of Corinthian Colleges, alleging the company boosted official placement rates by paying temporary employment agencies to hire students for brief stints after graduation (White 2015). Corinthian Colleges closed operations in 2016, adversely affecting more than 16,000 students in California and other states.

For-profit colleges are structured in ways that depart markedly from public and nonprofit schools. Rather than being fragmented and distributed, their decision structures are lean and centralized. Rather than attempting to please multiple stakeholders, they serve a unified set of shareholders. Curricular decisions are centralized at the corporate level and their "delivery model" is similarly centralized. Curriculum is highly structured, offering clear pathways to completion with little to no opportunity for exploratory or elective courses. The centralized model affords more efficiency in the creation, approval, and implementation of a new program. A process taking up to two years in a public college may be accomplished in a matter of weeks or months in a for-profit system.

Faculty are nontenured and exercise little to no discretion over curricular offerings or mode of instruction. Instructors receive training from college managers. A University of Phoenix informant reported that the college spent considerable effort and time in training and supporting new instructors. He stated that instructors underwent intensive training for about six months before teaching their first class and were continuously monitored and coached after the initial training period. According to this individual, all faculty were assessed twice a year with student evaluations being a key component of this process. The college thus enjoys much more latitude to remove underperforming instructors, or to replace an instructor with experience in one area with a different one should program priorities change.

For-profits have been quick to adopt online learning instruction. Many have a strong technology platform and have expanded much more rapidly than public colleges in offering online and/or blended courses. A respondent from the University of Phoenix reported the popularity of "flex-net" courses, where students are able to work at their own pace online (usually taking one course for six weeks) but come to campus to interact with faculty during scheduled hours to seek instructional support.

As reported, most of the for-profits also have the flexibility to meet students' needs, with courses offered in the evening, on weekends, as well as online. For-profit colleges can also adapt to change more easily than community or other public colleges because they own little to no property, foregoing the sunk costs of maintaining a campus with dedicated classrooms, dorms, student unions, or other amenities. Only a few, such as DeVry, operate (limited) student dorms. Many lease instructional space in shopping malls or in downtown commercial office buildings.

Yet another important advantage offered by for-profits is their willingness to present applicants with a generous interpretation of their previous academic accomplishments. While many public systems are strict in terms of how much academic credit entering students will be offered for their previous efforts in other schools, most for-profits not only grant credit for earlier courses taken, but also give credit for "life experiences" (e.g., employment and volunteer service) that count toward the pursuit of a degree.

Rosenbaum and colleagues (Rosenbaum, Deil-Amen, and Person 2006) find much to praise in many of these practices. In particular, they stress the value for student completion in colleges offering limited choices and highly structured "packaged deal plans" that lead to "an explicit career goal" (226). They also urge that colleges strengthen intake procedures for program choice, offer regular mandatory advising, and assist in managing financial and other types of problems affecting performance (230–31).

In spite of these substantial advantages, for-profits have had a difficult time gaining traction. As noted in case example 5.A, a decade of rapid growth around the turn of this century has been followed, after 2010, with sharp decreases in enrollments and, in some cases as described, failure of the company with severe implications for stranded students.

Concluding Comment

Most colleges in this country developed during the nineteenth century and adopted a structure consistent with managing the dependable production of a range of educational services. That structure was the bureaucracy, somewhat modified to provide discretion to academic subunits staffed by professionals. The pace of change was leisurely. Curricula was relatively standardized and the source of funding dependable. What competition existed focused on academic and athletic excellence and the status that accompanies such accomplishments. During the twentieth century, as demands for increased access grew for public colleges, state funding was the primary source of revenue, and state controls became more constraining.

But by the 1960s, colleges of all types were obliged to attempt to become strategic actors. They were increasingly subjected to market tests, competing for students and new sources of funding, assessed by media ratings and rankings. Leaders capable of strategic vision were sought after, and colleges were forced to become flexible in multiple ways: curricula change, enrollment policies, and timing and location of course offerings. Many also sought new ways to collaborate with companies, endeavoring to meet their demands for new kinds of courses and skill sets. Of course, colleges of the various types have been more and less successful in efforts to become increasingly agile and, for many of the public systems, most of the action has occurred in selected professional schools, such as engineering and business. Nonprofit colleges have, for the most part, opted to become specialized in currently popular vocational areas.

For-profit schools have benefitted from these conditions because their centralized and de-professionalized systems enable them to be responsive, and because they are subjected to fewer external controls. However, their interest in exploiting their advantages has led a number of these systems to overreach, overpromising results and understating requirements. In the absence of adequate controls—both internal professional and external regulative—for-profit colleges have yet to prove that they are eligible to be legitimate members of the portfolio of organizations providing quality higher education.

Policy Perspectives

MICHAEL W. KIRST, ANNE PODOLSKY,
LAUREL SIPES, AND W. RICHARD SCOTT

In our research, we have attempted to cast a wide net, capturing not only several decades of change but also a broad range of higher education programs and their supporting/constraining systems that have operated in the San Francisco Bay Area. Reaching even more broadly, we have endeavored to develop a regional approach to include important players in the Bay Area regional economy, in particular, employers and intermediate brokers, to help us understand how colleges connect to companies. In our work, we have discovered numerous and varied strengths as well as significant shortcomings in the performance of colleges and in their attempt to balance their commitment to academic values with their interest in serving economic/regional interests.

We would be remiss if we did not address the role of policy in these developments. Policies and administrative regulations at national, state, and regional levels have influenced the behavior of both colleges and companies, not to mention that of students and their families. Policies and the agencies that enforce them are the main tools available for public officials and private interests to directly attempt to influence educational systems and their programs. In this chapter, we offer our own ideas about possible positive changes that could be made as well as review some promising proposals that others have made for improving college performance.

The Limits of Policy Change

Before proceeding, however, we want to point out that policies are not the only external institutional force affecting the structure and functioning of colleges nor, indeed, are they the most powerful. Throughout this volume we have distinguished between regulative, normative, and cultural controls on organizations (see Scott 2014). The regulative elements prominently feature the mechanisms available to and utilized by policy makers: laws, authoritative oversight bodies, rules, and regulations. These elements are critical to the operation of modern society as we know it. Indeed, their prominence is one of the main features that distinguish modern from traditional societies. One of the most attractive features of regulative elements is that they are relatively "fast-moving" in comparison with norms, customs, conventions, and logics, which are "slow-moving" (Roland 2004). Policies are relatively easy to enact, to change, and to cancel. By comparison, normative elements, such as the academic standards enforced by professional associations, evolve much more slowly since they depend on deep commitments of many individuals acquired through lengthy socialization. And cultural elements such as customs and institutional logics change at an even slower pace. These include deep-seated views about, for example, the value of a college education and a liberal arts degree. As described in earlier chapters, change is underway in both academic norms and cultural beliefs. With the ascendency of neoliberal logics, the public good in the United States is increasingly defined in economic terms and educational value in market terms. These basic educational features are not determined by policy initiatives, although policies can reinforce them or attempt to limit their effects.

In short, policy proposals and programs have influence on educational programs, but they are not the sole or even the most important forces at work. Still, because they are among the most available and malleable of the mechanisms of control, we consider some of the more important ideas and arguments under consideration.

A Complex Terrain Posing Multiple Challenges

While the San Francisco Bay Area has a broad-ranging and deep array of postsecondary opportunities, serious problems threaten accessible, affordable, and high-quality educational opportunities and outcomes. These are evidenced by inadequate supply in academic domains with high student demand, and slow postsecondary adaptation to a rapidly changing regional economy. As the president of a Bay Area university said, "This economy changes exponentially, but our institution can only change incrementally."

Our study looks at the entire ecology of Bay Area postsecondary education over 40-plus years, including all types of institutions for all ages of pupils. Despite the existence of over 375 postsecondary education institutions in 2015, supply problems exist in a wide spectrum of programs, from liberal arts to technical education. The Bay Area demography has also changed as minority groups are drawn to the region. In addition, recent high school graduates and middle-aged adults struggle to find the educational options they want. This situation is particularly acute in the academic domains that are the foci of this study—business administration, information sciences, biology, and engineering. Moreover, since 2007, the San Francisco Bay Area, with 20 percent of the state's population, has generated well over half of the state's employment growth by adding 600,000 new jobs. Conversely, Los Angeles County, with 25 percent of the population, has produced less than 10 percent of the new jobs. In 2015, the Bay Area added 129,000 jobs, with 83 percent of the industrial sectors noting increases in the number of jobs (*Mercury News*, 2015).

The shortcoming of Bay Area postsecondary institutions to meet the increasing student demand is rooted in the tension between the two organizational fields, as highlighted throughout this volume. As chapter three describes, the traditional academic norms and structures were never designed to keep up with a fast-changing and rapidly growing regional economy like the Bay Area. Postsecondary values, organizational logics, standard operating procedures, faculty control, and institutional routines do not align well with much of what students need in one of the most dynamic economies in the world. Neither the state nor the region has a strategy to deal with this mismatch, or an understanding of the needed level of postsecondary investment to meet regional academic and economic demands.

Postsecondary governance and policy-making systems never envisioned an economy that continually reinvents itself. For example, the rapid growth and diversity of private postsecondary institutions was not embedded in the California Higher Education Master Plan of 1960. Moreover, only three comprehensive California State Universities (CSUs) exist in a region with over 7 million people. These universities are located in sections of the region with extreme traffic congestion and limited public transportation, which makes them hard for students to access. In California, three-fourths of the bachelor's degrees awarded annually come from CSU or UC. But there are seven postsecondary "systems" that operate in the Bay Area, and most have significant career and technical education components:

1. University of California
2. California State Universities
3. California community colleges
4. Private nonprofit
5. Private for-profit
6. Adult education run by K–12 schools
7. Workforce investment boards run by the state employment department

No state or regional entity attempts to coordinate or steer these seven systems. No one looks carefully at how quickly or deeply the seven separate components change to meet student desires for four-year degrees or workforce needs as the economy transforms. There is no mechanism at the state or regional level for all seven to even talk to one another, much less deliberate on shared issues. Developing an informed state policy and a regional coordinating entity is limited because of poor data systems that do not track much of the nongovernment sector (see appendix B).

The Bay Area's dynamic economy requires substantial adult reskilling, as well as a strong initial foundation in broad-based and general education. Some reskilling takes place in less formal and workplace settings, while initial education relies more heavily on traditional education organizations like colleges. Effective policies for these varied educational contexts will usually differ significantly. For example, policies for CSU, community colleges, and private liberal arts colleges probably will have a different approach than policies for nanodegrees, badges, and online education. State and regional leaders must be aware of the full spectrum of learning settings because demand is growing for all types of Bay Area education venues.

The three public systems contain many high-quality components but remain significantly separate from each other. Student transfer between them remains a work in progress with significant barriers. Most public colleges and universities use rolling three-year departmental program reviews to make changes in courses and other educational activities. But they struggle to keep up with changing student demand and an economy that requires new high-cost initiatives as Bay Area industry sectors grow and die. Bay Area high school graduates and working adults confront liberal arts majors and schools such as business administration and engineering that are declared "impacted" by one of the three public systems, which means that these academic domains cannot accept new or transfer students (see chapter four and case example 4.B). For example, San José and San Francisco State are impacted in every academic concentration explored in depth in this study. Obviously, state funding is part of the cause of this situation. However, other states, like New York, have avoided these issues, in part because of their integrated policy that takes into account private postsecondary education.

Private postsecondary institutions have filled some of the gaps in the Bay Area, including branch campuses of institutions such as Carnegie Mellon and Northeastern. The Bay Area relies more heavily on for-profit entities to educate students in the region, as compared to the national average. But these institutions are licensed and overseen by the California State Department of Consumer Affairs, which does not examine education quality. Many community colleges offer CTE (career and technical education), but community colleges were established to give priority to four-year transfer students. Often, the CTE division of community colleges is a separate entity from the transfer operation and is associated with less prestige and more adjunct faculty. It also costs more to offer many CTE courses compared to English or history because of the need to use current and expensive equipment. Despite the rapid growth in regional population, CTE public sector enrollment declined between 1992 and 2014. Some of this is because many young students prefer four-year degrees, but this should be offset by older students who seek CTE skills upgrading.

California's K–12 system—the base of the postsecondary system—ranks in the bottom 10 states on the National Assessment of Educational Progress (NAEP) in English and math. But high school students in California rank seventh in the country in the percentage of students who pass Advanced Placement exams. California has embraced and funded the college and career readiness approach of the Common Core, and this is supported by all California public postsecondary systems. This offers hope for the future. Moreover, California has funded new CTE career pathway programs that link secondary schools, community colleges, and employers.

The large array and diverse postsecondary institutions highlighted in this book require a nuanced and differentiated policy response that has a lifelong learning perspective. Policy makers cannot just change a few policies and expect to solve all the problems described in previous chapters. Moreover, improved policy must encompass federal, state, and local dimensions, and must surmount historical, structural, and political barriers. Compounding the challenges to implementing effective regional postsecondary policy is the lack of any models in California for how to do this. General education and liberal arts are crucial for student success in the Bay Area's diverse economy, and yet student demand for high-quality four-year degrees exceeds supply in the region.

Despite the many challenges presented throughout this book, California's postsecondary system has many successful attributes that hold promise in addressing the dynamics of Bay Area postsecondary supply and demand. There are many positive characteristics to build upon, such as California's relatively low tuition compared to most states and its above-state-average completion rate in public four-year institutions. University extension programs like UC Santa Cruz are well positioned to adapt to new technical occupational breakthroughs. Furthermore, California's private postsecondary institutions are often nimbler and can more quickly evolve according to the market context.

Common Core curriculum and college and career ready assessments will take time to transform students' preparedness for postsecondary education. California policy makers have embraced the Common Core because the curriculum and assessment are tightly linked to college and designed to lower remediation. Changes in K–12 alone, however, cannot solve all the major issues raised in this book about postsecondary performance. Consequently, California needs to consider sensible and innovative approaches to postsecondary education if it is to meet the needs of students and the Bay Area economy.

Below is an overview of some crucial obstacles that California faces in implementing more effective postsecondary policy. Following each overview is an illustration of various policy approaches that California, and the Bay Area, can take to alleviate these policy hurdles.

Evolution of California's Postsecondary State Policy, Governance, and Control

California's policy environment, which is the product of historical compromises and incremental decisions, has led to the stagnation of postsecondary outcomes. This patchwork quilt is still based on the 1960 Master Plan that was the envy of the nation for many years (see chapter four). However, the state steering role for all of California public and private postsecondary sectors has not been developed to meet current and future challenges. California is no longer an exemplar, as evidenced by California's 2012 degree-attainment rank placing the state twenty-third in the nation. In 2012, only 38.8 percent of California residents over 25 had an associate's degree or higher. While current degree completion is above the average for UC campuses, completion rates are well below the national average for CSU and community colleges. One potential reason for California's disappointing postsecondary outcomes is declining revenues for postsecondary institutions. California ranks toward the bottom in the nation in terms of postsecondary funding per student, even though student tuition and fees have increased.

The multiple causes of these outcomes for California impact Bay Area public postsecondary education similar to the rest of the state. The state steering role for all of California public and private postsecondary sectors is not well developed or able to meet the future challenge. The causes and consequences of California's lackluster postsecondary achievement have been well documented, as illustrated by the previous chapters.

A case study by Richardson and Martinez (2009: 70) comparing California's postsecondary outcomes to other states provides insight into the causes of the state's stagnant postsecondary outcomes. The study concluded that

> California is close to the median in effort, and its performance is mixed. On some measures it does very well; on others, it ranks among the lower-performing states. . . . One possible answer might rest with the rules for distributing state funds in California, which clearly favor the University of California at the expense of low-tuition community colleges. California's state constitution prohibits the direct appropriation of public funds to private institutions. The vaunted Master Plan (later described by its author as a treaty among the segments) discourages the use of market forces by tightly circumscribing the missions and enrollment pools for each of the three public sector systems. *A lack of consistent and effective statewide coordination has produced concurrent conditions where goals and priorities that can be established and implemented by a single segment can achieve very high levels of excellence, while those that require collaboration across the segments fall through the cracks* [emphasis added]. Shared faculty governance, along with other regulations mandated by the legislature, and taxpayer initiatives aimed at lowering property taxes have so muddied the waters for community colleges that distinguishing either their mission or their priorities can be a serious challenge. Moreover, the California legislature has separate committees for K–12 and postsecondary education that do not work together for intersegmental or regional concerns. California's flagging policy capacity is reflected in the nature of deliberation surrounding higher education. *Higher education policies continue to be focused on managing and regulating institutions more than on fundamental change to promote student success and economic development* [emphasis added].

Finney and colleagues (2014) reinforce the conclusions of Richardson and Martinez (2009) but focus on finance issues. These scholars note that each segment develops separate, individual policies based on the segment's goals, rather than collaborating to develop coordinating policies that focus on meeting California's economic needs. The authors argue that the lack of "a coordinated statewide finance policy . . . undermines the chances of improving degree attainment rates and threatens affordability." Finney and her coauthors (2014) also identify three significant weaknesses in California's postsecondary financial policy. The first is the state's failure to consider how appropriations, tuition, and financial aid align with statewide policies. In addition, the state does not tie appropriations to each segment's performance but, instead, often determines appropriations through annual compacts with the governor. Finally, the state's financial aid for college students is not meeting students' demand for postsecondary aid.

Richardson and Martinez's (2009) state case studies illuminate the stark differences between California and New York concerning the private sector, and suggest a future direction for California postsecondary policy. They find that "California does not view private institutions as part of the solution" in meeting student demand, even though a revision to the Master Plan in 1987 called the state to focus on the role of private institutions. The authors cite the lack of oversight across the sectors as the fundamental reason private institutions have not played a larger role in California's postsecondary policies. For example, the Cal Grant program provides limited money to students attending a private four-year college—$9,084 is the maximum stipend (in contrast to $12,240 for UC students). Interestingly, our study of the Bay Area emphasizes the important role of nonpublic postsecondary education (see chapters five and six).

In New York, Richardson and Martinez (2009) find that private institutions are a significant factor in New York's postsecondary planning. The New York State Board of Regents chartered the New York Commission on Independent Colleges and Universities (CICU) to "develop consensus among an extremely diverse group of [private, nonprofit] institutions and to influence higher education policy." The primary source of state support for private institutions is through the Tuition Assistance Program, which provides New York residents financial aid to attend approved public and private postsecondary institutions. In addition, public and private institutions collaborate to jointly offer thousands of degree programs. Richardson and Martinez also note that "the Board of Regents has also authorized four independent colleges to operate branch campuses or extensions centers on the campuses of public community colleges." The CICU in New York is one example of a state establishing a coordinating body to not only support and regulate postsecondary education but also to create coherent private postsecondary systems that meet the needs of students, institutions, and the state.

Fragmentation in Higher Education Governance and Policy Making

All of the seven California postsecondary systems confront several challenges, largely rooted in changing student demographics in a growing state. The most noticeable is the increased number of students seeking a postsecondary education in California (see chapter four). In addition, the average age of California's postsecondary students is no longer between 18 and 22 years but, instead, the average age of community college students is 30, and the average age of students graduating from California's four-year institutions is over 24 (Calisphere 2011). This means that many postsecondary students have full-time jobs and families to support. Consequently, these older students tend to want a "stripped-down version of education," meaning that they want convenient classes, advising and tutoring service, and high quality instruction (Calisphere 2011). These students also want low-cost classes, so they do not want to pay for services they do not need. Moreover, students are increasingly mobile and transfer between many types of institutions (Hossler et al. 2012). Perhaps most challenging is that the fastest growing demographic groups in the Bay Area, and throughout California, are those who have been least successful historically in college completions (Shulock, Moore, and Tan 2014).

The challenges created by a changing student body are heighted in community colleges. Many of the students who enter community colleges today fit the characteristics of those who are less likely to have access to college information and preparation (Kirst and Venezia 2004). Community colleges serve a large proportion of low-income, ethnic minority, and first-generation college students (Tinto 2004). Students from lower socioeconomic status levels and ethnic minority students are less likely to receive college counseling, be placed in college-preparation courses, or obtain information about college admissions and placement (Kirst and Venezia 2004). In turn, these factors contribute to students' unpreparedness for postsecondary education, increasing their risk of needing to enroll in remedial course work because of a lack of understanding of postsecondary standards and processes (see chapter four).

The failure of California's fragmented postsecondary system to respond to changing student demographics is illustrated by the challenges of transferring postsecondary credit hours in the state. Transferring student credit hours among the three public systems (UC, CSU, and community colleges) is difficult and often relies on decisions by individual academic departments in each system. While common course numbering and identification of content across the three systems is underway, it has been only partially implemented. A significant number of students are affected by the difficulties of transferring credit hours in such a fragmented system.

A 2012 study by the National Student Clearinghouse that included national postsecondary data found that one-third of all students transferred at least once within five years, with the majority of transfers occurring in students' second year, regardless of the direction of the transfer (i.e., vertical, lateral, or reverse), and a surprising number of students made their first transfer in their fourth and fifth years (Hossler et al. 2012). In the Bay Area, students move and rotate between postsecondary institutions for a variety of reasons, including their inability to get their priority classes at their home institution. For example, CSU students will attend a community college for a semester, and then return to CSU, but after that enroll at a different community college. These data demonstrate that many postsecondary students "swirl" between institutions and, consequently, are more likely to suffer from the lack of coordination among California educational institutions (Borden 2004). Similarly, CTE operates through several fragmented systems, restricted categorical funding, and diverse policy makers.

California's fragmented organization and control of postsecondary education has failed to improve postsecondary persistence rates, in part because it does not support the growing number of students who are moving across the separate systems. The California Master Plan is clearly outmoded for the dynamic needs of modern (and mobile) students. To improve student outcomes, institutions and policy makers critically need a detailed view of how students swirl between public and private systems and transfer rates to develop strategies and policies that facilitate successful student outcomes.

As noted, a major limitation of most higher education systems is a failure to recognize the large and growing number of working adults who are in mid-career but in need of further training and/or retraining. The pace of change in hi-tech industrial regions like Silicon Valley is such that the kinds of knowledge and skills that qualify a worker for a productive job as a new graduate are likely to become obsolete, requiring the acquisition of new and different kinds of competencies. Education and training programs cannot be limited just to those first entering the workforce, but must be reconstituted to support lifelong learning.

As chapter four indicates, the federal role in California is significant, particularly for student financial aid and research funding. However, the federal Higher Education Act has not been reauthorized since 2008. Our study did not focus on the federal role, but three primary concerns emerged. First, research suggests that federal student aid needs to be flexible. It does little to support older students, or those returning to postsecondary education. It is less aligned with students seeking credentials or specific job skills. Second, federal accreditation processes (i.e., the federal government's coordination of multiple nongovernmental regional entities responsible for accrediting institutions) need to be reexamined. Federal accreditation is fragmented, duplicative, and overly focused on processes and inputs. Accrediting is not well aligned with many components of skill training and some credentials. Employers need to play a larger role than in the past in revamping thinking about accreditation changes for adult education. Third, the federal reauthorization of the Higher Education Act needs to create more coherence with the Workforce Innovation and Opportunity Act, and the Perkins Career and Technical Education Act (New America Foundation 2015).

One promising development in California is the use of the new federal Workforce Innovation and Opportunity Act (WIOA) to formulate novel approaches to workforce preparation. One component of a state WIOA plan is reliance on new regional entities. Regional collaborations of public and private educational institutions, businesses, local governments, and other organizations can fill in gaps and devise contextual policies better than standard state approaches. But new California regional strategies are at an early stage, and it is too soon to determine whether they will be a solution. Indeed, the new bigger regions were defined in 2015. They are usually larger than community college districts or most traditional education entities like K–12 regional occupation programs and centers. Although this book is focused on the Bay Area region, the California postsecondary governance and policy system was never designed with a regional component in the mix.

In sum, the legacy of many decades of fragmented and incomplete policies leaves some basic components missing or unfinished, as described by a recent study from the Public Policy Institute of California (discussed in case example 7.A). This study identifies policies that may improve postsecondary attainment throughout California. Examples include increasing the number of students eligible to attend a UC or CSU, and expanding Cal Grants to cover more than tuition (e.g., housing, books) and increasing the amount of the grant for students at private institutions with good student outcomes.

CASE EXAMPLE 7.A
PUBLIC POLICY INSTITUTE OF CALIFORNIA POSTSECONDARY POLICY RECOMMENDATIONS

Our projections indicate that the demand for college graduates will outpace the supply by 2030, if current trends continue. The gap is substantial, with the economy needing 1.1 million more college graduates than the state will produce.

Closing the workforce skills gap will require improvements in several important areas. Here, we touch on a few key strategies for the state and its higher education institutions to pursue.

Increase Access

Students are more likely to earn a bachelor's degree if they first enroll in a four-year college, as opposed to a community college, even accounting for differences in academic preparation (Long, Kurlaender, and Grosz 2008; Johnson, Mejia, and Cook 2015). Therefore, one way to increase the number of college graduates in California is to increase the share of high school graduates eligible for UC and CSU. Doing so would also improve access for students from low-income and underrepresented groups.

Improve Completion and Time to Degree

At CSU, only 19 percent of students earn a bachelor's degree within four years, and just over half (54 percent) do so in six years; at UC, about 60 percent graduate in four years and 80 percent do so in six years. Both systems have used multiple strategies to increase these rates, including mandatory advising for at-risk students, eliminating bottlenecks by redesigning courses with high failure rates, increasing capacity for high-demand required courses, and using data to develop an early-warning system. These and other efforts should continue to be assessed to identify which are most effective. One new approach would be to provide fiscal incentives to colleges for increasing the share of students taking a full load (15 units). Much could be learned from the private, nonprofit colleges in the state, which have high four-year completion rates.

Expand Transfer Degrees

Because of the state's heavy reliance on community colleges, improving transfer pathways from those colleges to four-year institutions is essential. The vast majority of community college students do not earn a degree or certificate, leading to the large pool of California adults who have some college education but no degree. For new and current students, programs such as the associate degree for transfer should be expanded. These degrees guarantee access to CSU for community college students who fit the required criteria. However, these degrees still depend on individual agreements between specific campuses and specific majors. Expanding the program to include more majors and more campuses (including UC) should lead to increases in the number of students who transfer from the community colleges and ultimately earn bachelor's degrees.

Be Smart about Aid

Compared with other states, California has done a fairly good job at keeping college costs down, but more could be done. Grant and aid programs, including Cal Grants and institutional grants and scholarships, mean that most low-income, and even some middle-income, students do not have to pay tuition at the state's public colleges and universities. But other education costs are not well covered, and student debt has been rising—raising questions about whether Cal Grants should cover more than just tuition at public colleges. The state should also consider whether to increase the size of Cal Grants to students at private colleges with good graduation track records and whether to further decrease Cal Grants to institutions with low graduation rates and high loan-default rates. Improved evaluation of the effects of grant aid on student outcomes would help answer these questions.

Source: Johnson, Mejia, and Bohn 2016.

Community Colleges: An Illustration of California's Policy Issues

As the number of students attending community colleges has grown, community colleges have distanced themselves from secondary schools (see chapter four). For example, San José State started as a two year "normal school" for teacher preparation. It then became San José State College and now is a comprehensive university. Each transformation led to weaker linkage to K–12. After 1960, community colleges became the primary institutions for increasing college opportunity. Today, over 45 percent of undergraduates attend a community college, an increase of 10 percent in the last decade (Marcus 2005). This number has been increasing because of the significant use of community colleges in fast-growing states like California, Texas, and Florida. California, for example, enrolls two-thirds of its college freshmen into the community college system.

Originally, community colleges were funded like public schools with mostly local support, state supplements, and no tuition. In California, community colleges originated as part of the local K–12 system and were considered the thirteenth and fourteenth grades. For some students, however, the four-year systems dictated much of their curricula so as to facilitate transfer (Bracco, Callan, and Finney 1997). It was not until the 1950s that community colleges across the nation began to have their own governing boards.

The growth of community college enrollment was accompanied by a much-expanded mission and a loss of interaction with and focus on secondary education. The colleges expanded their mission to vocational education and community service. New and neglected populations beyond recent high school graduates were added, including displaced housewives, immigrants, older adults, and laid-off industrial workers. The comprehensive community college sent fewer and less clear signals to high school students about necessary academic preparation and skills needed to obtain vocational certificates. The impact of this detachment from secondary education has been profound, with many students entering community college unprepared for its demands.

About 70 percent of students entering California community colleges require remediation, a major risk factor for noncompletion of degree or certificate programs. Of all the English and math courses offered at the community college, 29 percent and 32 percent, respectively, are remedial (Cohen and Brawer 2003). The majority of the students enrolled in these remedial courses are of traditional college age and enter the college directly after high school. This implies that the high level of remediation is not just a result of having to refresh the skills of individuals who have been out of school for a while, but also of having to teach skills that were not received in high school. Increasingly, four-year institutions transfer their remediation to community colleges.

Existing CTE and Career Readiness Initiatives in K–12: Another Illustration of California's Fragmented Policies

K–12's role in preparing students for the workforce is a topic of longstanding tension. The issue raises big questions about the very purpose of public education (enlightenment or job training?), equity (who is getting prepared, and for what?), and agency (who gets to decide on a student's path, and at what point?). State law offers little guidance on these matters—just a few vague and scattered code sections that do not necessarily reflect current thinking. Education Code 51224.5, for example, states that school boards shall prescribe "separate courses of study," including those "designed to prepare prospective pupils for admission to state colleges and universities" and those "for career technical training"—a term that is not defined.

The state of California has previously developed plans for guiding CTE in high schools. There is a 2008 California State Plan for CTE that was required in order to qualify for the federal Perkins fund. The plan was jointly adopted by the State Board of Education and California Community Colleges Board of Governors, but does not reflect any of the education or workforce reforms initiated by the current governor or the current boards.

Although California's Local Control Funding Formula—the state's school funding formula enacted in 2013—has provided greater flexibility by eliminating most categorical programs, local districts still struggle to integrate a tangled array of funding streams into a coherent program plan at the school level. Many districts receive federal Perkins CTE funds for specific students and program activities. The state also continues to fund some categorical and competitive grant programs related to college and/or career readiness. Additionally, many districts are involved in privately funded initiatives aimed at improving career readiness, such as Linked Learning Alliance and a $500 million state program, Career Pathways Trust (CPT). CPT has much closer relationships among K–12, community colleges, and regional employers than most prior CTE programs. Thus, while the fiscal picture has been simplified somewhat at the state policy level, the task of weaving together these varied funding streams at the local level—each with its own goals, rules, metrics, and reporting processes—remains very complicated.

Even more difficult for K–12 districts is the work of building authentic partnerships with external entities in the postsecondary realm or the workforce. Traditionally, many K–12 districts have found it difficult to establish meaningful relationships with business leaders and local employers. Often such relationships—if they exist at all—are the result of one individual's connections and are not sustainable. This lack of capacity to partner across systems has been a major challenge in the creation of relevant career pathways programs that explicitly connect to post–high school opportunities. Sometimes there is not a clearly identified job-training program, community college degree or certificate program, or CSU major to "catch" students who are launched from high school CTE programs—or worse, such programs may exist but already are filled to capacity.

Over time, the layers of CTE policies and programs described above have led to entrenched organizational silos on the ground and, sometimes, a compliance mentality that focuses more on reporting requirements than holistic, long-term outcomes for students. In the long term, the state's move to local decision making may help districts develop strategic plans that set goals first and align spending second. The policy trend toward regionalism will encourage districts to initiate or strengthen relationships with postsecondary and workforce entities, but it will take time and likely more intentional assistance (both financial and technical) for many regions to become effective at leading themselves. Ideally, the education and workforce development entities in a region should have a shared understanding of their region's problems, goals, and how each member entity contributes to the regional plan. With this level of shared understanding, regional leaders can think creatively about how specific funds and partnerships can be leveraged to achieve the shared vision. The WIOA plan described above is a partial step in trying to drive this kind of regional cooperation from the state level.

Researchers have proposed many strategies for overcoming the weaknesses of California's CTE. In a personal interview, researcher Camille Esch suggests that state policy makers should encourage regional leaders in K–12 and postsecondary institutions to develop clear pathways between their respective CTE programs. Esch also encourages the state to invest in preparing individuals to teach CTE programs because finding teachers that can integrate CTE with core academic content is difficult (discussed in case example 7.B). Shulock and Moore (2013) recommend a statewide focus on community college policies (discussed in case example 7.C). These strategies, among others, provide a starting point for developing policies that improve students' outcomes in CTE.

CASE EXAMPLE 7.B
CAREER TECHNICAL EDUCATION POLICY RECOMMENDATIONS FOR
CALIFORNIA POLICY MAKERS

1. Incentivize regional capacity planning. In the rush to encourage career pathways, the state's K–12 education leaders should send a clear message that it is unacceptable to develop career pathways that lead students into dead ends. Pathway programs in high schools need to explicitly connect to specific postsecondary programs, and also need to take into account the capacity of those programs to accept the students who are graduating from K–12–based CTE programs. Because neither the local district nor the higher education entity can compel the other do anything differently, this is a place where a state accountability or fiscal mechanism might be needed.

2. Develop a strategy for preparing and providing professional development to CTE teachers. Most CTE reforms in the past few decades have aimed to better integrate CTE with core academic programming so that (1) CTE students are not isolated or unfairly tracked into predetermined outcomes, and (2) all students have opportunities to develop career-ready skills, whether they are headed to college first or into the workforce more quickly. However, a significant obstacle to integrating CTE with core academics is that few teachers are prepared to do it well. In many districts, CTE teachers and general education teachers likely need more professional development, as well as opportunities to collaborate together to develop shared goals, new strategies, and new courses.

Source: Esch 2015.

CASE EXAMPLE 7.C
RECOMMENDATIONS FOR STRENGTHENING CAREER
TECHNICAL EDUCATION

Increasing the number of students who pursue and earn certificates and associate degrees in California Community College career technical education (CTE) fields is an important component of the postsecondary completion challenge.

Strengthening CTE
- Ensure that high school and community college counselors have a better understanding of career pathways served by CTE programs. Shift

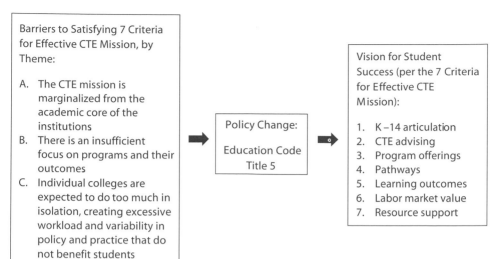

Figure CE7.C. Framework for policy reform to strengthen CTE.
Source: Shulock and Moore 2013.

colleges' focus from courses to programs. Ask colleges to offer roadmaps for progression through programs to help students achieve their goals.

- Improve processes for program approval, review, and discontinuation to ensure that low priority or outdated programs are phased out and energy and resources are directed to high-end, high-value programs.
- Develop pathways through which students may progress from short-term certificates into longer-term certificates and/or associate degree programs.
- Designate and provide capacity for an entity to be the primary provider of labor market information and analysis to the regional consortia and member colleges in order to keep programs responsive to market needs.
- Modify resource allocation to accommodate the high costs of many CTE programs and the essential nonteaching tasks of CTE faculty.
- Ensure stable, long-term funding for the emerging regional structures necessary to support regional collaboration across colleges.

Source: Shulock and Moore 2013.

Common Core: A Better Bridge to Postsecondary Education

One promising policy for reducing fragmentation among secondary and post-secondary California institutions is the Common Core State Standards (CCSS), which are designed to transform current instruction by focusing teacher attention on fewer, higher, and deeper standards that are aligned with postsecondary education. When implemented, they should reduce the need for college remediation. State assessment and accountability systems in California are being aligned with the Common Core's specific instructional approach. The deeper learning that CCSS aims to encourage has many elements, including:

- Understanding the meaning of ideas and their relevance to concrete problems
- Applying core concepts and modes of inquiry to complex, real-world tasks.
- Transferring knowledge and skills to new situations, to build on and use them.
- Communicating ideas and collaboration in problem solving.
- Reading increasingly complex texts closely and across content areas.
- Using evidence; interpreting with justification.
- Engaging in inquiry and research.
- Engaging in mathematical practices that use mathematical reasoning in application across content areas and being able to "understand," "describe," "explain," "justify," "prove," "derive," "assess," "illustrate," and "analyze," according to Common Core mathematics standards.

A promising development in California is the integration of K–12 Common Core and its Smarter Balanced Assessment with postsecondary systems. California administered the Smarter Balanced Assessment for the first time in 2015. It includes a scale score range for each student who reports college readiness, and includes the educational components students need to improve. The University of California is using CCSS for approving high school courses to satisfy admissions requirements. In addition, the SAT is revising its assessments with CCSS as a reference point. While more time is needed to integrate CCSS across educational segments, California has taken important steps to improve the articulation between K–12 and postsecondary education.

New Designs for Education and Workforce Improvement

Chapters five and six stress the difficulty that postsecondary institutions encounter in keeping up with student demand for more education in the fast-growing and changing sectors of the Bay Area. Improving this situation will require new concepts that modify the traditional degree structures and requirements and anticipate students' quickly evolving demands. The Bay Area has moved beyond the traditional notion of education first and employment later as the basic model. There is an increasing emphasis on competence rather than subjects. Some ideas worth considering are listed below (see also chapter three and case example 3.C).

Nanodegrees and Digital Badges

Nanodegrees and digital badges offer a narrow set of skills that can be clearly applied to a job, often through partnerships with employers. The degrees take between six to twelve months to complete online for as low as $200 a month. Some nanodegrees combine on-demand video lessons with online quizzes and longer projects. Some research suggests that this type of quick degree that can be directly applied to a career can improve the educational and occupational opportunities for low-income students (Porter 2014). However, quality control of these online degrees remains a concern (Porter 2014).

Stackable Degrees and Certificates

A portable, stackable degree includes the following characteristics: (1) students can earn shorter-term credentials with direct application to the labor market, and then build on the credential as they want to access advanced jobs and increased wages; (2) employers and educational institutions throughout the United States (and maybe world) trust the degree and credit transfer criteria or process; and (3) the degree/certificate/badge program is part of the California Career Pathways Trust, which includes a clear system and structure for developing the skills and credentials necessary for job entry through higher levels of employment (Austin et al. 2012). Stackable credentials have the potential to provide needed flexibility to students balancing work, school, and family, while simultaneously providing the ongoing learning that is the hallmark of most of the growing jobs in the economy. However, establishing clear educational and occupational standards to ensure that stackable credentials follow trajectories that lead to relevant training in an in-demand field will require updates to federal and state accountability regimes, as well as increased coordination between industry and education.

Specific Skill Bootcamps

Bootcamps are generally run by small, for-profit organizations, with courses running from two to four months. In the Bay Area, the most popular bootcamps are for individuals interested in learning how to code. One survey of coding bootcamps found that three-quarters of graduates were employed with average salaries of $76,000, representing a 44 percent increase from their salary prior to enrolling in the bootcamp (Lewin 2014). While this modern-day trade school has early signs of effectiveness, the key to enduring success will rest in the ability of the providers to maintain high quality in the face of market demands.

Elements of the "University of Everywhere"

Kevin Carey envisions a "University of Everywhere" that consists of online postsecondary courses open to the public, with digital lectures, digital readings, and virtual collaboration. This university will be inexpensive and meant for people to pursue lifelong learning (Carey 2015). The benefits and downsides to Carey's vision for postsecondary education overlap with those discussed above for nanodegrees, stackable credentials, and bootcamps (see also chapter four). However, elements of this mentality must be incorporated into postsecondary policy in order to respond to the exponential growth of knowledge that requires constant and continuous learning.

Lifelong Transcripts and Portfolios

The unbundled programs discussed above can be supported by two different records. The first is lifelong transcripts, which document an individual's education and related experiences (e.g., professional, volunteer) throughout the individual's life. These transcripts would be recorded online as "eTranscripts" and follow students throughout their academic and professional careers. A second record is an electronic portfolio, which can demonstrate a student's work and key skill and knowledge areas. These portfolios can provide evidence of students' achievement in authentic testing environments and, therefore, can be especially useful for prospective employers. Both of these recording mechanisms hold promise in meeting the needs of mobile students with vocational experience, as well as the needs of employers who want to better understand students' skills and knowledge.

Most of these concepts outlined above would be enhanced if employers, government, and educators provided more legitimacy and widespread recognition and quality assurance for them. Students need clearer signals about the potential worth of new alternative credentials. More employer coherence through recognition could help in the recruiting process. Social media can make alternative education and employment opportunities more visible to prospective students.

Applied Baccalaureate

An applied baccalaureate is a bachelor's degree that allows technical or occupational course work to be counted toward four-year degree completion. Applied baccalaureate degree curricula tend to use one of three models: (1) a career ladder model that allows students with technical associate degrees to complete additional technical course work, (2) a management model that provides business and administrative course work for students in the workforce who want to move into management positions, or (3) a general education model that accepts students from a variety of applied associate degree backgrounds and provides the general education courses required to meet state standards for baccalaureate degrees.

As of 2011, 39 states offered applied baccalaureate degrees, but not the CSU system. This type of degree could help increase the supply of students in Bay Area domains where there are capacity constraints in existing low-cost postsecondary programs such as engineering, business, and computer science. However, limited research exists on the effectiveness of this degree in meeting student and institution demands. Accordingly, Bay Area four-year educational institutions could pilot applied baccalaureate programs to determine whether they increase college completion rates, fill geographic gaps in baccalaureate access, and satisfy workforce demands in education, while not duplicating services provided by four-year institutions. The legislature authorized several community colleges to experiment with offering four-year applied degrees in community college by 2015. The tuition for these degrees is about $10,000. A few California community colleges began to experiment with four-year degrees in applied domains that CSU and UC were very unlikely to ever offer. The legislature restricted this four-year initiative to an experimental design to determine if a large-scale change needed to be made. Resistance from the four-year sectors is expected to be significant, and the enshrined legacy of the 50-year Master Plan will be difficult to overcome.

One study concluded that California needs to expand the enrollment at four-year institutions through the use of "hybrid" models, including applied baccalaureate programs (Geiser and Atkinson 2010). Rather than invest in creating new four-year institutions, the study suggests that collaborations between community colleges and state universities hold promise for increasing student degree attainment. For example, four-year institutions could have two-year branch campuses, where some community colleges are converted into satellites of state universities. This would eliminate the need for students to transfer between two- and four-year institutions. And the traditional divide between two- and four-year institutions would narrow, increasing the likelihood that students would move seamlessly from high school to baccalaureate programs to degrees.

Improving Accountability Measures and Mechanisms

In recent decades, the boundaries of the de facto field of higher education have expanded dramatically to incorporate a wide range of postsecondary institutions that provide a substantial proportion of higher educational services, particularly in the workforce arena (see chapter four). Many of these programs now qualify for federal student aid programs. This field expansion has occurred in the absence of accompanying effective oversight controls, whether from academic and professional associations or federal and state regulation. Indeed, as emphasized in earlier chapters, we lack relevant information about the most basic features of these programs: their curriculum, numbers and types of students served, numbers and qualifications of faculty, and internal governance mechanisms.

California also needs to examine new approaches to enhance postsecondary accountability. The focus should shift from college characteristics, process, time, and credits to a greater emphasis on learning and other professional outcomes. This transformation can be informed by career pathways, occupational clusters, and employer validation of education programs. Financial support and student aid must be linked to these new directions for both public and private education providers.

Online learning is a promising mechanism for expanding postsecondary opportunities but requires robust and thoughtful approaches to ensure that online programs benefit students. To make students more successful in online learning, a data-driven, integrated, and systematic approach is needed. Such learning should include faculty collaborating with "administrators, media developers, and information technology experts." In addition, online learning platforms should collect data to "better track student engagement . . . [and] customize learning" (Johnson, Mejia, and Cook 2015). Academic experts, as well as business and industry leaders, should inform the content of online courses. This will likely also help them understand the rigor of the course work, helping to ensure that such work is not discredited by the private sector.

The awarding of credits is another area in need of improved accountability. There is a wide array of possible credentials that include diplomas, occupational certificates and licenses, degrees, apprenticeship and specific skills certificates, and certifications within particular industries (see chapter three). A variety of entities award this complex array of degrees, certificates, and other credentials. These include, for example, employer-based training, industry-based certificates, union apprenticeships, and college associate degrees. In several occupations, a state-approved license is required before someone is permitted to work. In short, there is no common understanding of the labor market value or significance of most specific credentials. Some are earned by seat time and others by a performance assessment at the end of the program.

This complexity and confusion have led to calls for government deregulation and changes in accountability by Horn and Kelly (2015). Horn and Kelly suggest that higher education's reliance on "bundled" systems has influenced regulations, thereby stifling the development of low-cost "unbundled," "modular" courses (e.g., stackable degrees and certifications). In an approach embracing private sector solutions, the authors suggest three approaches to regulating unbundled programs. First, the federal government could provide aid to unbundled program providers that are transparent about their outcomes, costs, and student satisfaction. Second, federal policy makers could require new providers to use private capital to initially fund their programs, and then the federal government would reimburse (i.e., provide federal aid to) the new provider if they meet agreed-upon outcomes. Lastly, policy makers could leave the regulation of the new providers to the market and "let consumer demand and competition drive innovation"; however, as the authors note, this approach may be "slow and uncertain." Moreover, in our view, such approaches should be accompanied by stronger state regulations to curb the unethical and fraudulent claims made by some for-profit postsecondary programs. As noted, the Department of Consumer Affairs has proven itself to be incompetent in preventing these types of abuses, and its oversight powers need to be strengthened if these systems are to be included in the portfolio of legitimate educational providers.

State Performance Funding for Postsecondary Outcomes

In California, UC and CSU funding formulas evolved from old formulas built on student faculty ratios, class size, and student attrition rates. In addition, UC and CSU base funding is in part based on assumptions about the different resources needed for various disciplines and curriculum, as well as the flow of students both across and within institutions. For 40 years, annual state funding decisions have relied heavily upon additions to the prior-year base funding. In other words, the state has not significantly adjusted each institution's base funding, and instead has focused on short-term tactics to balance their budgets.

Recently, many states have begun implementing performance funding systems as a means to improve the performance and efficiency of their public higher education institutions. Performance funding shifts policy attention from inputs and regulations to outcomes. The early (and still present) type of performance funding takes the form of a bonus over and above regular state funding for higher education and is allocated on the basis of intermediate- and long-term indicators, including student persistence and completion. In newer approaches, performance funding is part and parcel of the regular state base funding allocation. Typically, the new funding formulas also take into account enrollments.

Performance funding has impacts on college finances, institutional knowledge of state priorities for higher education, and awareness of performance on state metrics. However, there is little evidence that performance funding increases state resources to improve institutions' capacity to respond to performance funding demands. Performance funding has led to intermediate institutional changes in the form of changes in academic and student services policies, programs, and practices that are intended to improve student outcomes. There is little evidence that performance funding programs significantly increase rates of student retention and graduation. Most careful quantitative analyses of the impacts of performance funding on retention/graduation have not found statistically significant impacts (Community College Research Center 2014).

Governor Jerry Brown and some legislators have supported the general ideas of performance funding, but no specific proposal has received serious consideration. The experience of other states suggests some ways to improve any California initiative using performance concepts:

1. Improve indicators and measures so as not to disadvantage community colleges. For instance, successful completion should be defined as including transfer to four-year colleges, and outcomes should be tracked over longer timeframes than just three years after college entrance.
2. Insulate performance funding from the state revenue cycle. Performance funding should represent a substantial portion of base funding.
3. Protect against possible unintended outcomes by removing disincentives to enrolling disadvantaged students and allow performance targets to vary based on student characteristics.

4. Protect academic standards by monitoring degree requirements and course grade distribution. To do this, educational institutions should take anonymous surveys of faculty to detect whether they are under pressure to weaken requirements, and by conducting student learning assessments.

5. Combat the narrowing of education missions by specifying indicators such as success in developmental education, general education, and continuing education (Community College Research Center 2014).

Data Limitations Hinder Good Policy Making

Historically, national data on higher education institutions have only included those schools providing students with federal financial aid. Broad-access colleges are less likely to take part in these programs and, as a result, are often ignored in these large, publicly available national datasets. To date, HEGIS (Higher Education General Information Survey) and IPEDS (Integrated Postsecondary Education System) are the most comprehensive data sources available for a longitudinal study like ours; however, we argue that results should be approached with caution, as exclusive use of these data insufficiently captures the complexity of the rapidly changing higher education sector. Thus, where possible, we have supplemented our analyses of HEGIS and IPEDS with information drawn from other directories.

Given the growing role of broad-access colleges, state and national agencies should consider collecting data on non-Title IV institutions and including them in their annual surveys. To provide a more complete picture of higher education, agencies may also wish to revise current practices regarding the reporting of data for branch campuses. In particular, for-profit colleges should be required to report data not simply at the system level but also by location of their branch campuses. Also, as corporate colleges have begun to play an increasing role in workforce training, ways must be found to include some information on the broad parameters of the training provided by companies. By excluding broad-access private institutions and campuses, these national datasets overlook a large swath of the postsecondary landscape serving students and contributing to regional economies. Lack of information on these types of colleges may unintentionally harm students, especially if these institutions fail to provide students with an adequate education and prepare them for labor market opportunities—as evidenced in the closures of many for-profit colleges. Finally, quantitative measures in these integrated national datasets ought to change over time and capture a wider array of information (e.g., gainful employment, online learning) that better reflects the needs of stakeholders (e.g., employers, parents, students).

The US system of higher education is rapidly evolving, and many items in federal datasets do not reflect the increasing heterogeneity of institutions. To better understand college-going behavior, it is important that the questions asked on surveys accurately reflect student choices and experience, market trends, and institutional policies and practices. Since California students swirl so much between postsecondary institutions and systems, there is a need for individual student tracking data. Appendix B provides details on data issues.

Political Challenges Inhibit Significant Policy Changes

Education reform in the United States in the last half-century has been overwhelmingly focused on K–12 education. Educators in primary and secondary schools have moved from a long history of relative autonomy to the point where the impact of local, state, and federal policy initiatives can be observed in almost every classroom in almost every public school in the country. In contrast, while higher education has undergone policy reforms, the impact of these reforms has not been nearly so far-reaching. Since public opinion has been shown to be a driver of policy change in other areas, we review trends in public opinion regarding how the public views K–12 education and higher education.

The evidence shows that the public has long been more critical of K–12 than higher education, and has only recently begun to question the value of a college education. K–12 education is now in an era where there are two main objectives: improving classroom instruction and increasing student achievement. Much of K–12 policy has shifted from being primarily concerned with adults, who are employees of school systems, to outcomes for children.

State funding for higher education appears to be on a downward trend. The federal government finds itself in the same situation, unable to keep up with the rapidly increasing costs of higher education. Instead, policy makers find themselves in the paradoxical position of valuing higher education more—due to the increased importance of a postsecondary degree—and being less able to directly control the system of higher education (Zumeta et al. 2012).

Our view is that most of the action regarding accountability for higher education has been in the form of what Tyack and Cuban term "policy talk" (Tyack and Cuban 1997). Public concern about—and disapproval of—K–12 education is much greater than concern about postsecondary education. For decades, the annual K–12 Gallup poll has given schools in a state or nation a C-, while a 2001 poll demonstrated that the public has given higher education a B/B+ (Immerwahr 2004; for more information about the PDK/Gallup poll, see http://pdkintl.org/programs-resources/poll/pdkgallup-poll-about). In the absence of an aroused public, postsecondary education reforms have not attracted much political momentum in the past 20 years. Generally, the public approves of higher education and does not demand reform. Instead, the public seeks greater access and affordability to colleges and universities (Immerwahr 1999b, 2000, 2004; Immerwahr and Foleno 2000; Immerwahr and Johnson 2007).

Research by John Immerwahr highlights key differences in public opinion regarding K–12 education and higher education. First, the public knows more about K–12 education and relatively little about higher education. Second, the public tends to view the quality of K–12 education as problematic, and higher education as being of very high quality. Third, the public generally understands that K–12 is paid for through tax dollars, while there does not appear to be broad public awareness that public colleges and universities also receive taxpayer support. Instead, most people think that higher education (even public higher education) is funded primarily by tuition. Given that most people don't know how these institutions of higher education work, there is unlikely to be consensus on policy changes for this sector.

One of the most important findings for the reform of higher education, particularly in the area of college success, has to do with the public's perceptions regarding responsibilities for educational success. Public Agenda (1999) reported that 75 percent of Americans say that almost all K–12 students can learn and succeed in school given enough help and attention. But for higher education, the story is quite different: "With virtual unanimity (91% to 7%) people think that the benefit of a college education depends on how much effort the student puts into it as opposed to the quality of the college that student is attending . . . when it comes to college, the public blames the problems on the [student] consumer, rather than on the producer" (Immerwahr 1999a: 10).

Many have characterized the political attitude toward higher education as being one of deference: higher education is funded to the best ability of policy makers, and more or less left alone. Compared with other areas of major state expenditure like K–12, transportation, and corrections, institutions of higher education and their leaders are rarely subject to scrutiny or micromanagement

(Zumeta 2001). Lobbyists also emphasize that higher education is so complex that only institutional leaders can oversee their own affairs.

The higher education subgovernments at the federal and state levels consist of three mutually reinforcing parts: the legislative subcommittees responsible for higher education, the bureaucracy and institutional governing boards responsible for implementing legislation, and the lobbying groups for higher education. The shape of policy making taking place in this subgovernment is characterized by mutual reinforcement and lack of conflict. In addition, higher education unions are much weaker than the dominant K–12 components of the National Education Association and American Federation of Teachers.

There are multiple possible events or changes that might move the problem of access to and success in higher education to the top of the policy-making agenda. These could include:

- A state funding crisis leading to the denial of admission to large numbers of students, particularly students from middle and upper-income families that have traditionally gone to college. Were this to occur, an outpouring of public anger would likely ensue.
- A lack of funding from the state level might not lead to denial of admission but rather to widespread cancellation of classes at public universities and colleges, meaning that many students were unable to graduate. The average time to graduation at bachelor's degree granting institutions could increase from six years to seven or eight, leading to anger from a broad swathe of middle- and upper-income voters.
- The generation gap in educational attainment is likely to widen. As the baby boomers retire, the lack of educational capital among the younger generation may become alarmingly clear and in many states rise to the level where it is considered a crisis. Pressure could come from the business community to "do something" about the lack of qualified candidates for jobs.
- The public could become aware of a drop in the quality of higher education at a time of rising tuition. K–12 reform efforts have been driven primarily by public concern about the quality of education. Several authors have documented what appears to be alarmingly low levels of student gains in higher education, and over time may garner more attention from policy makers (Arum and Roksa 2011; Doyle and Kirst 2015; Seifert et al. 2011).

The higher education community needs to come to greater clarity on the nature of the problems it faces related to students' educational progress, completion, and learning. What evidence is there that these problems can be solved with greater funding? With changes in curriculum? With changes in organizational structures? With changes in personnel? Although evidence is building, we know very little right now about both the nature of these problems and the kinds of interventions that would be most effective in increasing performance.

Regional Initiatives Needed to Improve Postsecondary Policy

San Francisco Bay Area leaders and community groups need to pay more attention and devise action plans to address the issues raised in this chapter and book, namely, the need for Bay Area postsecondary institutions to meet the demands of its dynamic economy while at the same time serving wider educational and civic needs. New federal and state policies can help but are unlikely to be sufficient to solve the major problems, or overcome the fundamental tension between the two organizational fields. Moreover, no Bay Area organization exists that can reasonably coordinate the 375 public and private postsecondary institutions in the region.

In the Bay Area, most business concerns have focused on issues in K–12 education, or on individual institutions of higher education. The bulk of regional philanthropy to improve policy has gone to K–12 and not postsecondary education. Despite the profound sub-regional differences in education and the economy highlighted in chapter five, there are no postsecondary sub-region organizations that span the full spectrum of postsecondary education. There are few forums that bring community groups, business, and higher education together to assess strengths and weaknesses of Bay Area postsecondary education, much less do anything to improve the coordination among these stakeholders.

The Bay Area needs a new regional entity to begin policy discussions and to coordinate among the region's education institutions. Such umbrella organizations exist in the Bay Area to coordinate environment, transportation, local government, and other domains. However, no similar organization exists for education. Below we describe two different approaches for developing a regional entity, and suggest a model for the Bay Area.

A State Centralized Plan for Regional Postsecondary Education

Shulock suggests for California a three-part strategy of regionalism, specialization, and technology, infused with collaborative mechanisms. She proposes establishing a regional consortium, which would include all public and private institutions, K–12, county offices of education, and regional employers to develop cross-sector strategies that increase college readiness and completions. The consortium would also define roles and responsibilities for its various institutions, align strategies to meet shared goals, develop meaningful pathways for students, and work with employers to provide work-based learning opportunities.

The three elements of regionalism, specialization, and technology would together help delineate roles for each institution and guide their growth, which can help ensure students are given access to comprehensive educational pathways. Shulock contends that it has been proven our current state-centric system lacks coherent direction and resources. Although certain region-based planning has happened periodically, it has lacked attention to systematic change.

All public higher education institutions would work together with their regional consortium to determine strategies for cost-effective education that align each institution with the regional budget. The state would develop coherent fiscal policies and perform the following functions: (1) adapt and require common, recognized definitions and accounting procedures like the Delta Cost Project; (2) collect and analyze finance data to translate regional needs to state-level budgets, including private institutions, other providers, and employers; (3) develop a coherent policy framework to ensure financial stability across the regions by using incentives to increase participation in financial aid that is provided on an equitable basis; (4) monitor the impact of fiscal incentives to ensure alignment with the goals of access, attainment, equity, affordability, and stability; and (5) monitor indicators of financial health, including state effort, affordability, and expenditures at the state, regional, and segment levels.

Accountability in this proposed model would be shared between higher education institutions, regional consortia, and the state. Each consortium would be accountable to the state for targets in college readiness and program specializations. The state would set performance goals for the segments and guide regional and state targets for performance. The state's accountability functions could include the following: (1) setting expectations for growth in degrees and certificates for each segment; (2) setting benchmarks for state-level performance on key indicators, including course completion rates, direct college entry rates, enrollment, and degrees awarded; (3) monitoring and potentially reassessing regional target-setting and annual outcomes to ensure that, collectively, regional targets meet state goals; (4) adjusting state policies to support regional plans for increased specialization and use of technology; and

(5) monitoring indicators of progress for meeting public agenda goals, which includes college readiness, degree attainment, student persistence rates, and job placement.

While comprehensive, Shulock's proposals have been criticized for featuring a top-down state role that centralizes control and prescription.

A Decentralized Plan for Regional Postsecondary Initiatives

In contrast to Shulock's top-down approach, the Board of Community College Governors created a bottom-up regional process: the Community College Regional Consortia. Eleven regional meetings focused upon needs assessments, financing expensive CTE courses, new curricula, and flexibility in hiring teachers. Governor Brown's 2016 budget provided a follow-up to this decentralized approach. Specifically, the governor provided $200 million of Proposition 98 general funds for the Strong Workforce Program, which would allow community colleges to offer more CTE courses. In addition, the Strong Workforce Program would implement a regional accountability structure and require that community colleges "collaborate regionally with their educational, workforce, labor, and civic partners" to create regional strategic plans that address regional education and workforce needs. Moreover, the strategic plans would inform the development, coordination, and availability of new and existing career technical education courses and programs.

The Community College Regional Consortia will identify regional industry clusters that are important to their economies, such as agriculture and advanced manufacturing. The sectors would be staffed with navigators who convene employers, administrators, and faculty to align resources and curriculum.

A decentralized approach like the Community College Regional Consortia seems more appropriate for California than Shulock's top-down statewide model. California is over 1,000 miles long and has over 39 million people. And the regional economies vary greatly as evidenced by the agricultural central valley compared to the Bay Area. Despite the promise of a decentralized approach, it is unclear how to expand the community college approach outlined above to include the UC, CSU, and private systems.

A Bay Area Regional Approach

The Bay Area needs a starting point and a new entity to begin policy discussions and action. Unfortunately, there are no existing California examples of this type of regional postsecondary approach, although regional leadership organizations like the Fresno Business Council, Silicon Valley Leadership Group (SVLG), and the Bay Area Council are possible initiators of a process. SVLG has an Industry/Higher Education Task Force to help shape and support the legislative efforts to increase the pipeline of employable graduates; and the Fresno Business Council was created in 1993 to "solve critical and systemic problems," including employment, education, and crime (Fresno Business Council 2013). The council was united by the belief that "unity in time, action, and leadership" is the only way to transform systemic and otherwise intractable public issues. Another example of regional coordination includes the California Economic Summit, which was first organized in 2012 to have "regional and state leaders develop a shared agenda to generate jobs and improve regional competitiveness" (California Economic Summit 2015a). The goals of this summit included "finding long-term workforce development funding for . . . career pathways to career technical education programs," as well as "expanding exposure to career options in manufacturing by connecting workforce training programs with manufacturers" (California Economic Summit 2015b).

Case example 7.D describes two regional agencies that span the Bay Area and have created integrated policy. The first agency, the Association of Bay Area Governments (ABAG) is a voluntary membership organization that includes elected officials from different levels of government, business leaders, and researchers who work together to address issues that impact their region. The second agency, the Metropolitan Transportation Commission (MTC) coordinates the creation, oversight, and financing of transportation throughout the nine counties in the San Francisco Bay Area. The Bay Area could serve as a model for other regions in California by adopting elements of the ABAG and MTC to create a regional education policy entity. Ideally, the regional entity would start organically, though the actions by regional leaders of traditionally disconnected organizations. Its members would include district superintendents, presidents of postsecondary institutions, business leaders, union members, and local politicians.

CASE EXAMPLE 7.D

DESCRIPTION OF TWO REGIONAL AGENCIES IN THE BAY AREA

The Association of Bay Area Governments (ABAG) is a voluntary membership and advisory organization with limited statutory authority, formed by a

Joint Powers Authority in 1961. The agency is governed by a General Assembly and Executive Board with standing and interagency committees. An elected official from each of the nine counties and 101 member cities and towns serve as delegates to its General Assembly, which determines policy annually, adopts the annual budget and work program, and reviews the policy actions of ABAG's Executive Board. ABAG convenes its General Assembly once every year in April. Conference attendees include Bay Area elected officials, civic and business leaders, and researchers from leading academic institutions who identify and address policy issues that impact the entire region. The 38-member Executive Board meets bimonthly to make operating decisions, appoint committee members, authorize expenditures, and recommend policy (ABAG 2015a).

The Metropolitan Transportation Commission (MTC) is the transportation planning, coordinating, and financing agency for the nine-county San Francisco Bay Area, created by the state legislature in 1970. Over the years, the agency's scope has grown, and it is now three agencies in one, functioning as MTC as well as the Bay Area Toll Authority (BATA) and the Service Authority for Freeways and Expressways (SAFE) (ABAG 2015b). The Commission's work is guided by a policy board of 21 individuals, with 18 of the commissioners designated as voting members. Sixteen of the voting commissioners are appointed by local elected officials in each county (Metropolitan Transportation Commission 2015).

As an example of an effective innovative improvement, the MTC created the "clipper card," which allows riders to use all types of transit systems—trains, buses, and trolleys—over the entire area by just pressing the card against an electronic device. Other examples of useful reform implemented are the many trails and bike paths that crisscross the Bay Area region, ignoring city and county boundaries. These innovations demonstrate the upside potential of regional collaboration.

The regional education entity would establish a strategic vision for K–16 education in the region, and work to coordinate the necessary stakeholders to enact this vision. It would also develop a comprehensive online resource for students to understand the variety of opportunities in the entire ecology of the over 375 Bay Area postsecondary education institutions, as well as less formal training centers such as bootcamps. This resource base would list the enrollments and openings in each educational program in real time, as well as the credentials students receive for completion of the educational program, ranging from badge to graduate degree. Moreover, the resource base would be designed to serve individuals earning their initial high school or postsecondary degree, as well as adults looking to reskill.

With this resource, prospective students would know the requirements to enter and exit. Student opportunities to transfer between institutions would be specified, including bottlenecks like CSU program impaction, where no more students can transfer. Student aid opportunities would also be specified for each program. Students would know about the size of the queue for over-enrolled programs like nursing at a particular institution. And they would know how to apply to multiple institutions that fit their ambitions. In addition, prospective students could review data on each institution's outcomes, including completion and employment. This comprehensive Bay Area resource would provide a unique regionally focused student resource as compared to resources that are currently provided by the state systems, individual institutions, and popular media.

The development of this regional resource would also provide the necessary information to develop effective policy advocacy tailored to Bay Area context and needs. The resource's data could inform a variety of policy domains, including K–12 preparation, demand/supply imbalances, effective educational approaches, data needs, finance, and performance incentives.

In sum, the regional coordinating functions would encompass at least six domains:

1. Connections to employers about current and emerging workforce needs, and feedback to institutions about the adequacy of student preparation.
2. Interinstitutional coordination around supply/demand, to take advantage of opportunities for consolidation of programs that do not have the critical mass necessary to provide a full range of programs at each individual campus (this should build on institutional program reviews; see report on program reviews by Ris on the Gardner Center website).
3. Data warehousing and analysis on key performance indicators relevant to student readiness and transitions from K–12 to postsecondary education, including progression to certificates/degrees. This function could also allow those interested to connect to communities of expertise in data analytics and benchmarking and communication,

including connections to national datasets that would be difficult for individual institutions to do on their own.

4. Goal setting and collaboration across multiple institutions. The regional entity should be bringing players together to talk in fairly specific terms about future needs, goals for access and attainment including attainment gaps, and what meeting those needs and goals means for the institutions within the region. This would include an analysis of the potential capacity for growth in the private sector.

5. Connections to other regional entities across the state to periodically meet to talk about what works, what does not, and to identify emerging "best practices," particularly around remediation and transfer.

6. Discussion of how state policies could be better tailored to regional needs such as charging different tuition in high- and low-cost regions. The higher-cost regions would then get more student aid per pupil.

Regional entities should have varied roles that begin with the meeting and convening of stakeholders, experts, and the public to discuss and understand the current status and needs of Bay Area higher education. This volume makes a start on understanding but does not provide a regional action agenda. More regional data are needed to inform local, state, and national audiences about the Bay Area context, situation, needs, and alternative strategies/policies.

Concluding Remarks—The Need for a New Bay Area Policy Agenda

This chapter makes a case for postsecondary education to reach the very earliest stage of policy—getting a prominent place on the policy agenda. The policy cycle begins with an agenda-setting process that identifies problems that require government attention, decides which issues deserve the most attention, and defines the nature of the problem. Many issues contend for scarce time and attention on an overloaded federal, state, and local policy agenda. Some problems fester for a long time and then explode onto the top of the agenda. Many issues languish in the background and never attain a prominent place in the deliberations of public and private policy-making bodies. So far, regional higher education problems have taken a back seat to K–12 education, transportation, housing, and environmental issues. We argue that this should not continue.

The beginning of creating a policy agenda is to realize the magnitude and importance of the problem. This volume has provided ample evidence that the

Bay Area postsecondary problems are acute and will not be solved by a continuation of current policies, such as reliance on the state Master Plan for higher education. The initial stage of policy challenges assumptions about the nature of the problem, and how to begin to solve it. We have provided some insights on how to address the problem, but more needs to be done to create specific policy options, as well as investigating their strengths and weaknesses.

The initial stage of policy formulation also creates a policy frame for how to deal with it. A policy frame is a narrative about a troublesome situation that often uses metaphors. For example, estate taxes can be framed as a "death tax" or equity issue based upon inherited privilege. Each narrative policy frame contains a different view of reality and a special way of conceptualizing the problem (Stone 2002). A slum can be seen as a blighted area to be torn down, or a place to be redesigned and rebuilt. We have framed the Bay Area higher education scene as one that has many strengths, including an assortment of varied and often strong colleges and universities, but also problematic dimensions, including inequitable basic education opportunities, labor shortages, a source of inefficient businesses, and colleges that cannot keep up with changing demands. Our policy frame relies on the tension between the organizational cultures discussed in chapter three.

It often depends on the insights and efforts of "policy entrepreneurs" to secure a prominent place on the policy agenda by seizing an issue, framing it, and advocating for it. A whole book has been written on how charter K–12 schools went from a nascent idea to a major national education intervention (Mintrom 2000). Currently, there is no entrepreneur for either the domain of regional higher education, or the particular Bay Area context. We hope this book will galvanize the attention and subsequent action to create a high priority agenda based in part upon the issues raised by the data and analysis presented in this volume.

Appendix A

EXPERT INFORMANTS

Interviewee	Role/Title
Emily Allen	Associate Dean of the College of Engineering at San José State University
Steve Barley	Associate Chair of the Department of Management Science and Engineering at Stanford University
Dennis Cima	Senior Vice President, Silicon Valley Leadership Group
Michael Cubbin	President of Bay Area Metro Region at DeVry University
Bernadine Fong	Former President of Foothill College
Jim Gibbons	Former Dean of Stanford University's School of Engineering (1984–96)
Doug Henton	Chairman and CEO, Collaborative Economics, Inc.
Dennis Jaehne	Associate Vice President of Curriculum and Undergraduate Studies at San José State University
William M. Kays	Former Dean of Stanford University's School of Engineering (1972–84)
Steve Levy	Director, Center for Continuing Study of the California Economy
William F. Miller	Former Vice President and Provost of Stanford University (1971–78)
Judy Miner	President of Foothill College
Al Moye	Former Director of university affairs at Hewlett Packard
Brian Murphy	President of De Anza College
AnnaLee Saxenian	Dean of the School of Information, University of California, Berkeley
Maureen Sharberg	Associate Vice President of Student Academic Services at San José State University
Nancy Shulock	Former Director of the Institute for Higher Education Leadership and Policy at Sacramento State University
Van Ton-Quinlivan	Vice Chancellor of Workforce and Economic Development; formerly with Heald College
Barbara Waugh	Former Recruiter for Hewlett Packard and Hewlett Packard Labs

Appendix B

INVISIBLE COLLEGES: THE ABSENCE OF BROAD-ACCESS
INSTITUTIONS IN NATIONAL POSTSECONDARY DATASETS

BRIAN HOLZMAN

Our research draws on data from the Higher Education General Information Survey (HEGIS) and the Integrated Postsecondary Education System (IPEDS). Both HEGIS and IPEDS are national surveys of higher education institutions fielded by the National Center for Education Statistics (NCES) of the US Department of Education. For most research in higher education, HEGIS and IPEDS are extremely useful databases since they capture critical information for nearly all of the accredited, degree-granting, Title IV institutions in the county. However, when it comes to studying broad-access colleges such as nonaccredited, for-profit institutions that offer short-term training programs, HEGIS and IPEDS have important limitations.

Higher Education General Information Survey (HEGIS)

HEGIS, which predated IPEDS, was established after the 1965 Higher Education Act and was conducted from 1965 to 1987. HEGIS requested institutions participating in the Title IV federal student financial aid program to complete an annual survey (Jaquette and Parra 2014). The survey collected annual data on institutional characteristics (e.g., level, control), fall enrollment, earned degrees, employment, and finance. Other institutional information was also gathered but less systematically (e.g., libraries, in-state student residency). The HEGIS sample included "institutions of higher education," which were defined by NCES as being "accredited at the college level by an agency or association recognized by the Secretary, U.S. Department of Education. These schools offered at least a one-year program of study creditable toward a degree and they were eligible for participation in Title IV Federal financial aid programs" (National Center for Education Statistics, Institute of Education Sciences, US Department of Education 2014b).

Limitations of HEGIS

Although HEGIS strived to be a central catalog of US postsecondary institutions, broad-access colleges were often overlooked. For instance, descriptive analysis of institutions in California show that in 2010, for-profit colleges granted about 59 percent of the state's "short-term certificates"—awards lasting less than one academic year (Jez 2012). HEGIS disproportionately excludes for-profit colleges and universities because students typically pursue programs in these institutions (e.g., occupational or technical certificate programs) that do not lead to a two- or four-year degree. Further, many broad-access colleges, particularly private proprietary institutions, are nonaccredited and may not offer federal student financial aid. Our inventory and map of San Francisco Bay Area colleges and universities on the Gardner Center website (http://gardnercenter.stanford.edu) show that many broad-access colleges are excluded in these large national datasets.

There is also the problem of missing data. With the exception of three years of data available on the IPEDS website (1980–81, 1984–85, and 1985–86), HEGIS is no longer maintained by NCES. Although the Inter-University Consortium for Political and Social Research (ICPSR) at the University of Michigan holds a large catalog of HEGIS files, several years and survey components are missing. With the help of an NCES staff member, we were able to obtain additional data discs for our research. Although these sources allowed us to assemble institutional data for most years between 1970 and 1986, there are still several gaps. For example, data are unavailable for the following school years: 1971–72, 1975–76, 1976–77, 1979–80, and 1981–82. A number of components are also unavailable, including information about earned degrees in 1970. Based on our communication with NCES, we learned that data were frequently lost because of repeated format conversions—from magnetic tapes to files on diskette to records on the internet. Therefore, we were constrained in our ability to construct a full 40-year panel dataset with complete information.

The Integrated Postsecondary Education System (IPEDS)

IPEDS replaced HEGIS and was phased in between 1985 and 1989 (Fuller 2011). Currently, IPEDS fields nine survey components: institutional characteristics, completions, 12-month enrollment, student financial aid, fall enrollment, finance, 150 percent graduation rates, 200 percent graduation rates, and human resources (Jaquette and Parra 2014). Data collected in IPEDS are similar to those in HEGIS, thus supporting longitudinal panel data analysis.

Initially, IPEDS included Title IV and non-Title IV institutions, regardless of degree-granting or accreditation status (Jaquette and Parra 2014). These types of institutions were included because IPEDS replaced two other national higher education surveys in addition to HEGIS: the Survey of Non-Collegiate Postsecondary Institutions and the Vocational Education Data System. After 2001, IPEDS only included Title IV institutions; non Title IV institutions are allowed to take part in the IPEDS surveys, but their participation is voluntary. Trend analyses compiled by NCES that use HEGIS and IPEDS typically limit their samples to degree-granting, Title IV institutions for cross-time comparability (Kena et al. 2014). Analyses of non-Title IV institutions must use other sources, such as data from state education agencies (Cellini and Goldin 2012).

Limitations of IPEDS

Exclusive use of IPEDS to examine broad-access colleges, such as proprietary institutions, is problematic. At a national level, Jez (2014) finds that IPEDS excludes approximately half of for-profit colleges and universities and 27 percent of students attending such institutions. The for-profit institutions that participate in the Title IV program also charge significantly higher tuition rates and may not represent the average proprietary school. In addition, while the inclusion of non-degree-granting colleges and universities is an improvement from HEGIS, those captured in IPEDS likely comprise a small sample of the total number—since many institutions offering certificates do not participate in federal student financial aid programs.

How institutions report information to IPEDS also varies. While some colleges and universities report a main campus and each branch campus separately, others report the main campus only, combining branch campus enrollments and completions (Jaquette and Parra 2014). This way of aggregated reporting appears to be the case with DeVry University and the University of Phoenix. Our inventory of broad-access colleges in the region shows that DeVry has four locations: Daly City, Fremont, San Jose, and San Francisco; however, none of these locations are listed in IPEDS. In fact, the only location in California that DeVry reports in IPEDS is in Pomona—an area outside the scope of our analysis (National Center for Education Statistics, Institute of Education Sciences, US Department of Education 2014a). Likewise, while the San Jose campus of the University of Phoenix is captured in IPEDS, the Livermore and Oakland branches are not.

Other types of branch campuses not reported separately to IPEDS include extensions of four-year public universities, which focus on continuing education for working adults (e.g., UC Berkeley Extension). Also overlooked are remote graduate programs at private universities, including Babson College in San Francisco and Carnegie Mellon University in the Silicon Valley. It is unclear the degree to which these inconsistencies in branch reporting influence trends in college enrollment and completion, particularly in a study focused on broad-access colleges. Regardless, the exclusion of some branch campuses in IPEDS, due to nonsystematic data reporting, results in an incomplete representation of the postsecondary ecology.

Measurement-Related Issues

The measures generated by NCES also have shortcomings. For instance, in recent years, NCES has produced files reporting 150 percent and 200 percent graduation rates. Yet these data do not disaggregate the types of degrees. An excerpt from a prominent education blog points out that this is akin to comparing apples and oranges: "Comparing the graduation rates of community colleges and for-profit colleges is effectively judging the rate at which students earn associate degrees against shorter certificate programs. This apples-to-oranges comparison is particularly problematic since one would of course expect the completion rate of shorter programs to be higher since there are fewer opportunities to drop out and no need to re-enroll for a second academic year" (Miller 2014).

Another measurement-related issue pertains to the data on fall enrollment. If a student matriculates in a certificate program in the spring or summer and completes his or her credential, he or she will not be counted in the institution's fall enrollment records. While IPEDS does produce a 12-month unduplicated head count, this file is a relatively new addition to the survey and cannot be examined in historical analyses spanning multiple decades.

In a report focused on using IPEDS to examine community college student outcomes, Offenstein and Shulock (2009) highlight further limitations. The authors observe that NCES gives institutions significant flexibility in identifying degree-seeking students. While some colleges and universities may survey students on their personal goals, others may construct a behavioral measure that determines a degree-seeking student by the number of credits accumulated or the number of particular courses completed. This type of flexibility complicates efforts to compare institutions, particularly broad-access colleges, which typically have more students who "enroll for personal enrichment, to take adult education courses, to take a small set of courses to learn a specific skill, or to take advantage of the low cost and convenience of community colleges to 'try out' college" (Offenstein and Shulock 2009: 7). Moreover, these researchers note that while measurement error may occur when developing survey or behavioral indicators of degree intent, the survey-based measure could be problematic given that students likely overestimate their abilities. As a result, the measure may more accurately reflect students' degree aspirations rather than more realistic expectations.

IPEDS is further flawed in that it does not accurately reflect the variety of student attendance patterns in higher education institutions, particularly at broad-access colleges. The graduation and transfer-out rates in IPEDS only include full-time students who enter in the summer or fall terms. While this may be acceptable for four-year colleges and universities, data show that less than half of two-year college students began their postsecondary education by attending full time in the summer or fall (Offenstein and Shulock 2009). Subsequently, the graduation rates reported in IPEDS may be inaccurate.

Three rates are reported: 100 percent, 150 percent, and 200 percent of normal time to completion. A 200 percent completion rate, for instance, allows four-year college students eight years to complete a bachelor's degree and two-year college students four years to complete an associate's degree. This may appear to be a generous time allotment, but given that many students—particularly at community colleges and other broad-access institutions—attend part time or require remedial education, the window may in fact be too short. Also, the time limits pose unique challenges when studying students seeking certificates. That is, if a certificate typically takes six months to complete, a 200 percent graduation rate should only consider a certificate completed if a student finishes within a year. However, it is unclear whether institutions are abiding by this rule or including all certificates completed within two, three, or four years (Offenstein and Shulock 2009).

Offenstein and Shulock also observe shortcomings in IPEDS regarding student mobility. Students attending broad-access colleges are quite mobile and may attend multiple institutions before they receive a degree (Offenstein and Shulock 2009). Yet, these student attendance patterns are not tracked across institutions. Consequently, if students transfer to another community college to finish their education, they may be recorded as a dropout from the initial institution, regardless of whether they finish their degree at the new school. Another issue pertains to how student transfers are reported in IPEDS. Transfers to two- and four-year institutions are treated similarly and not disaggregated. One of the key functions of community college systems is to facilitate four-year college attendance, and differences between horizontal and vertical transfers should not be ignored. There is also a problem with counting transfers. Namely, some students may transfer to a four-year institution but yet complete only a certificate or associate degree. However, institutions often treat transfer as a secondary outcome and solely report the degree attained. Such an approach underestimates the number of vertical transfers, obscuring the transfer function of broad-access colleges (Offenstein and Shulock 2009).

Response Rates to HEGIS and IPEDS

Despite the limitations with HEGIS and IPEDS, response rates to these surveys remain high. According to NCES staff, both surveys have had increasing participation. Since 2002, IPEDS has been mandatory for Title IV institutions, and the survey has achieved a nearly 100 percent response rate (Ginder, Kelly-Reid, and Mann 2014). Failure to complete the survey may result in a $35,000 fine or removal of Title IV status and federal student financial aid eligibility (National Center for Education Statistics, Institute of Education Sciences, US Department of Education 2014c). Institutions that do not participate in Title IV programs may still complete the surveys, and an incentive for doing so is inclusion in the higher education directories such the *Basic Student Charges* or *College Navigator* (National Center for Education Statistics, Institute of Education Sciences, US Department of Education 2014a). The *Basic Student Charges* directory was distributed to higher education institutions during the HEGIS period. The *College Navigator* is a website listing information from IPEDS on US colleges and universities.

Conclusion

Our research on the role of broad-access colleges in the San Francisco Bay Area represents a small and imperfect snapshot of the changing postsecondary landscape—since much of the existing information on higher education institutions is limited. Historically, national data on higher education institutions have only included those schools providing students with federal financial aid. Broad-access colleges are less likely to take part in these programs and, as a result, are often ignored in these large, publicly available national datasets. To date, HEGIS and IPEDS are the most comprehensive data sources available for a longitudinal study like ours; however, we argue that results should be approached with caution, as exclusive use of these data insufficiently captures the complexity of the rapidly changing higher education sector. Thus, we have supplemented our analyses of HEGIS and IPEDS with information drawn from other directories. Still, our inventory of colleges and universities in the San Francisco Bay Area is also not without its flaws. While we use many additional sources, it is likely that we have not captured all broad-access institutions. To correct this, future studies may want to consider using computer programming languages (e.g., Python) to extract listings from www.YP.com or other web-based telephone and mailing address directories to ensure that all types of postsecondary training institutions have been recorded.

Given the growing role of broad-access colleges, state and national agencies may want to consider collecting data on non-Title IV institutions and including them in their annual surveys. To provide a more complete picture of higher education, agencies may also wish to revise current practices regarding the reporting of data for branch campuses. By excluding broad-access institutions and campuses, these national datasets overlook a large swath of the postsecondary landscape serving students and contributing to regional economies. Lack of information on these types of colleges may unintentionally harm students, especially if these institutions fail to provide students with an adequate education and prepare them for labor market opportunities—as evidenced in the closures of many for-profit colleges. Finally, quantitative measures in these integrated national datasets ought to change over time and capture a wider array of information (e.g., gainful employment, online learning) that better reflects the needs of stakeholders (e.g., employers, parents, students). The US system of higher education is rapidly evolving, and many items in federal datasets do not reflect the increasing heterogeneity of institutions. To better understand college-going moving forward, it is important that the questions asked on surveys more accurately reflect student behaviors, market trends, and institutional policies and practices.

References

Abbott, Andrew. 2001. *Chaos of Disciplines*. Chicago: University of Chicago Press.
———. 2002. "The Disciplines and the Future." In *The Future of the City of Intellect: The Changing American University*, edited by Steven Brint, 206–30. Stanford, CA: Stanford University Press.

Acclaim. 2014. *Open Badges for Higher Education*. Accessed November 20, 2016. www .pearsoned.com/wp-content/uploads/OPen-Badges-for-Higher-Education.pdf.

Adams, Caralee. 2015. "One-Third of College Students Transfer Schools before Graduation." *Education Week*, July 8. Accessed December 14, 2016. http://blogs .edweek.org/edweek/college_bound/2015/07/_of_money_a_college_students_parents _make_does_correlate_with_what_that_person_studies_kids_from_low.html.

Ahrne, Göran, and Nils Brunsson. 2008. *Meta-Organizations*. Northampton, MA: Edward Elgar.

Alcorn, Brandon, Gayle Christensen, and Ezekiel J. Emanuel. 2014. "Who Takes MOOCs?" *The New Republic*. Accessed December 14, 2016. https://newrepublic.com /article/116013/mooc-student-survey-who-enrolls-online-education.

Alexander, Lamar. 2012. *Higher Education Accreditation Concepts and Proposals*. Senate Committee on Health, Education, Labor, and Pensions. Accessed December 14, 2016. http://www.help.senate.gov/imo/media/Accreditation.pdf.

Alliance for Excellent Education. 2013. *Expanding Education and Workforce Opportunities through Digital Badges*. Accessed December 14, 2016. http://all4ed.org/reports -factsheets/expanding-education-and-workforce-opportunities-through-digital -badges.

Alstete, Jeffrey W. 2004. *Accreditation Matters: Achieving Academic Recognition and Renewal*. Hoboken, NJ: Wiley Periodicals.

American Council on Education. 2012. *Assuring Academic Quality in the 21st Century: Self-Regulation in a New Era*. American Council on Education National Task Force on Institutional Accreditation. Accessed December 14, 2016. https://www.acenet.edu /news-room/Documents/Accreditation-TaskForce-revised-070512.pdf.

Armstrong, Elizabeth A., and Laura T. Hamilton. 2013. *Paying for the Party: How College Maintains Inequality*. Cambridge, MA: Harvard University Press.

Arns, Robert G., and William Poland. 1980. "Changing the University through Program Review." *Journal of Higher Education* 51:268–84.

Arum, Richard, and Josipa Roksa. 2011. *Academically Adrift: Limited Learning on College Campuses*. Chicago: University of Chicago Press.
———. 2015. "Measuring College Performance." In *Remaking College: The Changing Ecology of Higher Education*, edited by M. W. Kirst and M. L. Stevens, 169–89. Stanford, CA: Stanford University Press.

Ash, Katie. 2012. "'Digital Badges' Would Represent Students' Skill Acquisition." *Education Week*, June 13. Accessed December 14, 2016. http://www.edweek.org/dd /articles/2012/06/13/03badges.h05.html.

Asimov, Nanette. 2015. "CCSF Wins Reprieve: Shutdown Averted with 2-Year Extension." *SF Chronicle*, Accessed December 14, 2016. http://www.sfgate.com/bayarea /article/SF-City-College-shutdown-averted-with-two-year-6015600.php.

Association of Bay Area Governments (ABAG). 2015a. "ABAG Governance." Accessed December 14, 2016. http://abag.ca.gov/overview/governance.html.

———. 2015b. "Partner Agencies." Accessed December 14, 2016. http://abag.ca.gov /overview/partners.html.

Astin, Alexander W., and Anthony L. Antonio. 2012. *Assessment for Excellence: The Philosophy and Practice of Assessment and Evaluation in Higher Education*. Lanham, MD: Rowman and Littlefield.

Aud, S., W. Hussar, W. Johnson, F. Kena, G. Roth, E. Manning, and J. Zhang. 2012. *The Condition of Education 2012* (NCES 2012-045). Washington, DC: US Department of Education, National Center for Education Statistics.

Auerhahn, Louise, Bob Brownstein, Cindy Chavez, and Esha Menon. 2012. *Saving the Middle Class: Lessons from Silicon Valley*. Working Partnership USA. Accessed December 14, 2016. http://www.wpusa.org/Publication/LIVE2012-online.pdf.

Austin, James T., Gail O. Mellow, Mitch Rosin, and Marlene Seltzer. 2012. *Portable, Stackable Credentials: A New Education Model for Industry-Specific Career Pathways*. New York: McGraw-Hill Research Foundation. Accessed December 14, 2016. http://www.jff.org/sites/default/files/publications/materials/Portable Stackable Credentials.pdf.

Backes, Ben, and Erin Dunlop Velez. 2014. *Who Transfers and Where Do They Go? Community College Students in Florida*. CALDER, American Institutes for Research. http://www.aefpweb.org/sites/default/files/webform/39th/backes_velez _commcollege_aefp2014.pdf. Site discontinued.

Bailey, Thomas R., Norena Badway, and Patricia J. Gumport. 2002. *For-Profit Higher Education and Community Colleges*. Stanford, CA: National Center for Post-Secondary Improvement, Stanford University. Accessed December 14, 2016. http://web.stanford .edu/group/ncpi/documents/pdfs/forprofitandcc.pdf.

Bale, Rachael. 2013. "Protestors Demand Help for City College of San Francisco." *KQED*, March 14. Accessed December 14, 2016. http://ww2.kqed.org/news/2013/03 /14/protestors-demand-city-hall-help-fund-san-francisco-city-college.

Barak, Robert J. 1982. *Program Review in Higher Education: Within and Without. An NCHEMS Executive Overview*. Boulder, CO: National Center for Higher Education Management Systems. Accessed December 14, 2016. http://files.eric.ed.gov/fulltext /ED246829.pdf.

Barley, Stephen R., and Gideon Kunda. 2004. *Gurus, Hired Guns, and Warm Bodies: Itinerant Experts in a Knowledge Economy*. Princeton, NJ: Princeton University Press.

Bastedo, Michael N., and Nicholas A. Bowman. 2009. "U.S. News & World Report College Rankings: Modeling Institutional Effects on Organizational Reputation." *American Journal of Education* 116 (2): 163–83. doi: 10.1086/649437.

Bay Area Council Economic Institute. 2014. "UC Berkeley: Stimulating Entrepreneurship in the Bay Area and Nationwide." San Francisco: Bay Area Council Economic Institute.

Bear, Charla, and Jon Brooks. 2013. "Dept. of Education Ruling Won't Solve CCSF Accreditation Woes." *KQED*, August 14. Accessed December 14, 2016. http://ww2 .kqed.org/news/2013/08/13/106731/city-college-ccsf-accreditation.

Becker, Gary S. 1964. *Human Capital: A Theoretical and Empirical Analysis, with Special Reference to Education*. New York: Columbia University Press.

Benner, Chris. 2002. *Work in the New Economy: Flexible Labor Markets in Silicon Valley*. Oxford: Blackwell.

Benner, Chris, Laura Leete, and Manuel Pastor. 2007. *Staircases or Treadmills: Labor Market Intermediaries and Economic Opportunity in a Changing Economy*. New York: Russell Sage Foundation.

Benner, Katie. 2015. "Obama, Immigration and Silicon Valley." *Bloomberg View*, January 22. Accessed December 14, 2016. http://www.bloombergview.com/articles /2015-01-22/obama-immigration-reform-h-b1-visas-and-silicon-valley.

Bergeron, David A. 2013. "Applying for Title IV Eligibility for Direct Assessment (Competency-Based) Programs." *Inside Higher Ed*, March 19. Accessed December 14, 2016. https://www.insidehighered.com/sites/default/server_files/files/FINAL GEN 13-10 Comp Based 3-14-13 (2).pdf.

Berman, Elizabeth Popp. 2012. *Creating the Market University: How Academic Science Became an Economic Engine*. Princeton, NJ: Princeton University Press.

Berman, Elizabeth Popp, and Catherine Paradeise, eds. 2016. *The University under Pressure*. Bingley, UK: Emerald Group.

Blau, Peter M. 1970. "A Formal Theory of Differentiation in Organizations." *American Journal of Sociology* 63:58–69. Accessed December 14, 2016. http://www.jstor.org /stable/2093199.

———. 1973. *The Organization of Academic Work*. New York: John Wiley and Sons.

Bledstein, Burton J. 1976. *The Culture of Professionalism: The Middle Class and the Development of Higher Education in America*. New York: W. W. Norton.

Blivin, Jamai, and Merriliea J. Mayo. 2013. "The New World of Work from Education to Employment." *Innovation Intake*, May 5–11. Accessed December 14, 2016. http://issuu.com/theinnovationintake/docs/intake_may_2013/1?e=4882245/2273321.

Bloland, Harland G. 2001. *Creating the Council for Higher Education Accreditation*. Phoenix: American Council on Education/Oryx Press.

Blue Sky Consulting Group. 2015. *Background Paper: Funding Career and Technical Education (CTE) Programs at California Community Colleges*. California Community Colleges Taskforce on Workforce. Accessed December 14, 2016. http://doingwhat matters.cccco.edu/portals/6/docs/SW/CTE Funding Report for CCCCO REVISED 20150420.pdf.

Blumenstyk, Goldie. 2011. "U. of Phoenix Hit With New Whistle-Blower Lawsuit over Recruiting Practices." *Chronicle of Higher Education*, May 27. Accessed December 14, 2016. http://chronicle.com/article/U-of-Phoenix-Hit-With-New/127714.

———. 2015. "U. of Phoenix Looks to Shrink Itself with New Admissions Requirements and Deep Cuts." *Chronicle of Higher Education*, June 30. Accessed December 14, 2016. http://chronicle.com/article/U-of-Phoenix-Looks-to-Shrink/231247.

Bogue, E. Grady, and Kimberly Bingham Hall. 2003. *Quality and Accountability in Higher Education: Improving Policy, Enhancing Performance*. Westport, CT: Praeger.

Borden, Victor M. H. 2004. "Accommodating Student Swirl: When Traditional Students Are No Longer the Tradition." *Change* 36 (2): 10–17. Accessed December 14, 2016. http://eric.ed.gov/?q=two+state+solution+for+Israel&ff1=audResearchers &ff2=autBorden%2c+Victor+M.+H.&id=EJ701873.

Bound, John, Michael F. Lovenheim, and Sarah Turner. 2010. *Increasing Time to Baccalaureate Degree in the United States*. Cambridge, MA: National Bureau of Economic Research.

Bourdieu, Pierre. 1971. "Systems of Education and Systems of Thought." In *Knowledge and Control: New Directions for the Sociology of Education*, edited by M. Young, 189–207. London: Collier-Macmillan.

———. 1977. *Outline of a Theory of Practice*. Cambridge: Cambridge University Press.

Boyd, Aaron. 2014. "New Site to Bolster Cybersecurity Community, Workforce." *Federal Times*, December 22. Accessed December 14, 2016. http://www.federaltimes.com/story/government/cybersecurity/2014/12/22/new-site-cybersecurity-community-workforce/20774373.

Bracco, Kathy Reeves, Patrick M. Callan, and Joni E. Finney. 1997. *Public and Private Financing of Higher Education: Shaping Public Policy for the Future*. Westport, CT: American Council on Education/Oryx Press.

Bradley, Gwendolyn. 2004. "Contingent Faculty and the New Academic Labor System." *Academe* 90 (1): 28–31. doi: 10.2307/40252585.

Breneman, David W. 2006. "The University of Phoenix: Icon of For-Profit Higher Education." In *Earnings from Learning: The Rise of For-Profit Universities*, edited by D. W. Breneman, B. Pusser, and S. E. Turner, 70–73. Albany: SUNY Press.

Breneman, David W., Brian Pusser, and Sarah E. Turner, ed. 2006. *Earnings from Learning: The Rise of for-Profit Universities*. Albany: SUNY Press.

Brint, Steven, and Jerome Karabel. 1989. *The Diverted Dream: Community Colleges and the Promise of Educational Opportunity in America, 1900–1985*. New York: Oxford University Press.

———. 1991. "Institutional Origins and Transformations: The Case of American Community Colleges." In *The New Institutionalism in Organizational Analysis*, edited by W. W. Powell and P. J. DiMaggio, 37–60. Chicago: University of Chicago Press.

Brint, Steven. 2002. "The Rise of the 'Practical Arts.'" In *The Future of the City of Intellect: The Changing American University*, edited by Steven Brint, 231–59. Stanford, CA: Stanford University Press.

Brint, Steven, Mark Riddle, and Robert A. Hanneman. 2006. "Reference Sets, Identities, and Aspirations in a Complex Organizational Field: The Case of American Four-Year Colleges and Universities." *Sociology of Education* 79 (3): 229–52. Accessed December 14, 2016. http://www.jstor.org/stable/25054315.

Brittingham, Barbara. 2009. "Accreditation in the United States: How Did We Get to Where We Are?" *New Directions for Higher Education* 145:7–27. doi: 10.1002/he.331.

Brunsson, Nils, and Bengt Jacobsson. 2000. *A World of Standards*. Oxford: Oxford University Press.

Burke, Lindsey M., and Stuart M. Butler. 2012. *Accreditation: Removing the Barrier to Higher Education Reform*. The Heritage Foundation. Accessed December 14, 2016. http://files.eric.ed.gov/fulltext/ED535877.pdf.

California Economic Summit. 2015a. "The Goal." Accessed December 14, 2016. http://www.caeconomy.org/pages/the-goal.

———. 2015b. "Workforce and Workplace." Accessed December 14, 2016. http://www.caeconomy.org/pages/progress-entry/workforce-workplace.

California State Department of Education. 1960. *A Master Plan for Higher Education in California 1960–1975*. Sacramento. Accessed December 14, 2016. http://www.ucop.edu/acadinit/mastplan/MasterPlan1960.pdf.

California State University. 2002. *The California State University Enrollment Management Policy and Practices*. Accessed December 14, 2016. https://www.calstate.edu/acadres/docs/CSU_Enroll_Mngment_Policy_Practices.pdf.

————. 2015. "Mission of Course Redesign with Technology." *Course Redesign with Technology.* Accessed December 14, 2016. http://courseredesign.csuprojects.org/wp /mission.

California Student Aid Commission (CSAC). 2016. "Eligibility Criteria for Cal Grants 2016." Accessed December 14, 2016. http://www.cesac.ca.gov.

Calisphere. 2011. *A Guarantee of Equity for Older, Part-Time Students.* University of California. Accessed December 14, 2016. http://content.cdlib.org/view?docId =hb538nb32g;NAAN=13030&doc.view=frames&chunk.id=div00017&toc.id =0&brand=calisphere.

Callan, Patrick M. 2009. *California Higher Education, the Master Plan, and the Erosion of College Opportunity.* National Center for Public Policy and Higher Education. Accessed December 14, 2016. http://www.highereducation.org/reports/cal_highered/cal _highered.pdf.

————. 2014. "Higher Education in California: Rise and Fall." In *Higher Education in the American West,* edited by L. F. Goodchild, R. W. Jonsen, P. Limerick, and D. A. Longanecker, 233–56. New York: Palgrave Macmillan.

Campaign for College Opportunity. 2015. *Access Denied: Rising Selectivity at California's Public Universities.* Accessed December 14, 2016. http://collegecampaign.org /portfolio/november-2015-access-denied-rising-selectivity-at-californias-public -universities.

Cappelli, Peter. 2001. "Assessing the Decline of Internal Labor Markets." In *Sourcebook of Labor Markets: Evolving Structures and Processes,* edited by I. Berg and A. L. Kalleberg. New York: Kluwer.

Carey, Kevin. 2015. *The End of College: Creating the Future of Learning and the University of Everywhere.* New York: Riverhead Books.

Carmody, Tim. 2011. "Without Jobs as CEO, Who Speaks for the Arts at Apple?" *Wired* (August). Accessed December 14, 2016. http://www.wired.com/2011/08/apple -liberal-arts.

The Carnegie Classification of Institutions of Higher Education. 2015. "Definitions." Indiana University School of Education. Accessed December 14, 2016. http:// carnegieclassifications.iu.edu/definitions.php.

Carnegie Commission on Higher Education. 1973. *The Purposes and Performance of Higher Education in the United States.* New York: McGraw-Hill.

Carnevale, Anthony P. 1993. *The Learning Enterprise.* Washington, DC: US Government Printing Office.

Carnevale, Anthony P., and Donna M. Desrochers. 2001. *Help Wanted . . . Credentials Required: Community Colleges in the Knowledge Economy.* Annapolis Junction, MD: Community College Press.

Carnevale, Anthony P., and Stephen J. Rose. 2011. *The Undereducated American.* Georgetown University Center on Education and the Workforce. Accessed December 14, 2016. http://files.eric.ed.gov/fulltext/ED524302.pdf.

Carnevale, Dan. 2006. "Rule Change May Spark Online Boom for Colleges." *Chronicle of Higher Education,* February 3. Accessed December 14, 2016. http://chronicle.com /article/Rule-Change-May-Spark-Online/14648.

Carroll, Constance M. 2010. "Evaluation Report: City College of San Francisco. A Report Prepared for the Accrediting Commission for Community and Junior Colleges Western Association of Schools and Colleges." Accessed December 14, 2016. ccsf.edu/Offices/Research_Planning/pdf/2006-06_WASC_Report.pdf.

Casner-Lotto, Jill, and Linda Barrington. 2006. *Are They Really Ready to Work? Employers' Perspectives on the Basic Knowledge and Applied Skills of New Entrants to the 21st Century U.S. Workforce.* The Conference Board, Corporate Voices for Working Families, Partnership for 21st Century Skills, Society for Human Resource Management. Accessed December 14, 2016. http://files.eric.ed.gov/fulltext/ED519465.pdf.

Castilla, Emilio J., Hokyu Hwang, Ellen Granovetter, and Mark Granovetter. 2000. "Social Networks in Silicon Valley." In *The Silicon Valley Edge*, edited by C-M. Lee, W. F. Miller, M. Cong Hancock, and H. S. Rowen, 218–47. Stanford, CA: Stanford University Press.

CB Insights. 2015. "Venture Capital Database." Accessed December 14, 2016. https://www.cbinsights.com.

Cellini, Stephanie Riegg, and Claudia Goldin. 2012. "Does Federal Student Aid Raise Tuition? New Evidence on For-Profit Colleges." *American Economic Journal: Economic Policy* 6 (4): 174–206. doi: 10.1257/pol.6.4.174.

Chandler, Alfred D., Jr. 1977. *The Visible Hand: The Managerial Revolution in American Business.* Cambridge, MA: Belknap Press.

Child, John. 2005. *Organization: Contemporary Principles and Practice.* Oxford: Blackwell.

Clark, Burton R. 1970. *The Distinctive College: Antioch, Reed, and Swarthmore.* Chicago: Aldine.

———. 1983. *The Higher Education System: Academic Organization in Cross-National Perspective.* Berkeley: University of California Press.

———. 1985. *The School and the University: An International Perspective.* Berkeley: University of California Press.

———. 1998. *Creating Entrepreneurial Universities: Organizational Pathways of Transformation.* New York: Pergamon Press.

Clawson, Dan. 2009. "Tenure and the Future of the University." *Science* 324 (5931): 1147–48. doi: 10.1126/science.1172995.

Cohen, Arthur M., and Florence B. Brawer. 2003. *The American Community College.* San Francisco: Jossey-Bass.

Cohen, Patricia, and Chad Bray. 2016. "University of Phoenix Owner, Apollo Education Group, Will Be Taken Private." *New York Times*, February 8. Accessed December 14, 2016. http://www.nytimes.com/2016/02/09/business/dealbook/apollo-education-group-university-of-phoenix-owner-to-be-taken-private.html?_r=0.

Cole, Jonathan R. 2010. *The Great American University: Its Rise to Preeminence, Its Indispensable National Role and Why It Must Be Protected.* New York: Public Affairs.

College of Alameda. 2016. "ATLAS." Accessed December 14, 2016. http://alameda.peralta.edu/atlas.

Colyvas, Jeannette A., and Walter W. Powell. 2006. "Roads to Institutionalization: The Remaking of Boundaries between Public and Private Science." *Research in Organizational Behavior* 27: 305–53. doi: 10.1016/S0191-3085(06)27008-4.

Community College Research Center. 2014. *Performance Funding: Impacts, Obstacles, and Unintended Outcomes.* Teachers College, Columbia University. Accessed December 14, 2016. http://ccrc.tc.columbia.edu/media/k2/attachments/performance-funding-impacts-obstacles-unintended-outcomes-2.pdf.

Conrad, Clifton F., and Richard F. Wilson. 1985. *"Academic Program Reviews: Institutional Approaches, Expectations, and Controversies. ASHE-ERIC Higher Education Report No. 5, 1985."* Association for the Study of Higher Education. Washington, DC: Institute of Education Sciences.

Cook, Constance Ewing. 1998. *Looking for Higher Education: How Colleges and Universities Influence Federal Policy*. Nashville: Vanderbilt University Press.

Council for Higher Education Accreditation (CHEA). 2006. "Presidential Perspectives on Accreditation: A Report of the CHEA Presidents Project." Washington, DC. Accessed December 14, 2016. http://files.eric.ed.gov/fulltext/ED494267.pdf.

———. 2008. *CHEA Almanac of External Quality Review*. "Fact Sheet #1: Profile of Accreditation." Washington, DC: Council for Higher Education Accreditation. http://www.chea.org/pdf/fact_sheet_1_profile.pdf#search=%22almanac of external quality review 2008%22.

Craig, Douglas B. 2000. *Fireside Politics: Radio and Political Culture in the United States, 1920–1940*. Baltimore: Johns Hopkins University Press.

Craven, Eugene C. 1980. "Evaluating Program Performance." In *Improving Academic Management*, edited by P. Jedamus and M. W. Peterson, 432–57. San Francisco: Jossey-Bass.

Curtis, F. Philler. 1984. *Menlo School and College: A History*. Atherton, CA: Mayfield Publishing.

Dasgupta, Partha, and Paul David. 1994. "Toward a New Economics of Science." *Research Policy* 23:487–521. doi: 10.1016/0048-7333(94)01002-1.

De Anza College. 2015a. "CAD Department." https://www.deanza.edu/cdi. Site discontinued.

———. 2015b. "Careers in Design and Manufacturing Technologies." *Workforce Education*. Accessed December 14, 2016. https://www.deanza.edu/workforceed/manf-cnc.html.

Deil-Amen, Regina. 2015. "The 'Traditional' College Student: A Smaller and Smaller Minority and Its Implications for Diversity and Access Institutions." In *Remaking College: The Changing Ecology of Higher Education*, edited by M. Stevens and M. Kirst, 134–68. Stanford, CA: Stanford University Press.

Delta Cost Project. 2013. "Delta Cost Data." American Institute for Research. Accessed December 14, 2016. deltacostproject.org/delta-cost-data.

Deming, David, and Susan Dynarski. 2010. "Into College, Out of Poverty? Policies to Increase the Postsecondary Attainment of the Poor." In *Targeting Investments in Children: Fighting Poverty When Resources Are Limited*, edited by P. B. Levine and D. J. Zimmerman, 283–302. Chicago: University of Chicago Press.

Deming, David, Claudia Goldin, and Lawrence Katz. 2013. "For-Profit Colleges." *The Future of Children* 23 (1): 137–63. Accessed December 14, 2016. http://futureofchildren.org/futureofchildren/publications/docs/23_01_07.pdf.

Desrosier, James. 2010. "UCSC Extension in Silicon Valley: Early Enrollment Incentives." *Continuing Higher Education Review* 74:142–49. Accessed December 14, 2016. http://files.eric.ed.gov/fulltext/EJ907258.pdf.

Deterding, Sebastian. 2011. "Situated Motivational Affordances of Game Elements: A Conceptual Model." *CHI* (April): 3–6. doi: ACM 978-1-4503-0268-5/11/05.

DiMaggio, Paul J., and Walter W. Powell. 1983. "The Iron Cage Revisited: Institutional Isomorphism and Collective Rationality in Organizational Fields." *American Sociological Review* 48 (2): 147–60. Accessed December 14, 2016. http://www.jstor.org/stable/2095101.

Donabedian, Avedis. 1966. "Evaluating the Quality of Medical Care." *Milbank Memorial Fund Quarterly* 44 (3.2): 166–203. doi: 10.111.1/j.1468-0009.2005.00397.x.

Douglass, John Aubrey. 2010. *From Chaos to Order and Back?: A Revisionist Reflection on the California Master Plan for Higher Education@50 and Thoughts about Its Future*.

Berkeley, CA: Center for Studies in Higher Education. Accessed December 14, 2016. http://eric.ed.gov/PDFS/ED511964.pdf.

———. 2011. "Can We Save the College Dream?" *Boom: A Journal of California* 1 (2): 25–42. doi: 10.1525/boom.2011.1.2.25.

Doyle, William R., and Michael W. Kirst. 2015. "Explaining Policy Changes in K–12 and Higher Education." In *Remaking College: The Changing Ecology of Higher Education*, edited by M. W. Kirst and M. Stevens, 190–213. Stanford, CA: Stanford University Press.

Dudnick, Laura. 2015. "CCSF to Meet with Accrediting Commission over Possible Reversal of 2013 Decision." *San Francisco Examiner,* July 6. Accessed December 14, 2016. http://www.sfexaminer.com/ccsf-to-meet-with-accrediting-commission-over -possible-reversal-of-2013-decision.

Duncan, Arne, Secretary of Education. 2011. "Digital Badges for Learning." US Department of Education. Accessed December 14, 2016. http://www.ed.gov/news /speeches/digital-badges-learning.

Eaton, Judith S. 2009. "Accreditation in the United States." *New Direction for Higher Education* (145): 79–86. doi: 10.1002/he.337.

The Economist. 2014a. "Creative Destruction." *Economist,* June 28. Accessed December 14, 2016. http://www.economist.com/news/leaders/21605906-cost-crisis -changing-labour-markets-and-new-technology-will-turn-old-institution-its.

The Economist. 2014b. "Is College Worth It?" *Economist,* April 24. Accessed December 14, 2016. http://www.economist.com/news/united-states/21600131-too-many -degrees-are-waste-money-return-higher-education-would-be-much-better.

The Economist. 2014c. "Workers of the World, Log in." *Economist,* August 16, 1–7. Accessed December 14, 2016. http://www.economist.com/news/business/21612191-social -network-has-already-shaken-up-way-professionals-are-hired-its-ambitions-go-far.

The Economist. 2015a. "Excellence v. Equity." *Economist,* March 28, 3–19. Accessed December 14, 2016. http://www.economist.com/news/special-report/21646985 -american-model-higher-education-spreading-it-good-producing-excellence.

The Economist. 2015b. "Keeping It on the Company Campus." *Economist,* May 14, 55–56. Accessed December 14, 2016. http://www.economist.com/news/business/21651217 -more-firms-have-set-up-their-own-corporate-universities-they-have-become-less -willing-pay.

Eisgruber, Christopher L. 2011. "Letter to Susan D. Phillips at State University of New York at Albany." May 26. Accessed December 14, 2016. http://www2.ed.gov/about /bdscomm/list/naciqi-dir/2011-spring/naciqi-6-2011-comments.pdf.

Elliot, Leslie Orrin. 1937. *Stanford University: The First Twenty-Five Years.* Stanford, CA: Stanford University Press.

Elsner, Paul A., and George R. Boggs. 2005. *Encouraging Civility as a Community College Leader.* Lanham, MD: Rowman and Littlefield.

Emslie, Alex. 2013. "City College of San Francisco Enrollment Plunges after Threat-ened Accreditation Loss." *KQED,* July 25. Accessed December 14, 2016. http://ww2 .kqed.org/news/2013/07/25/104635/ccsf-accreditation.

Esch, Camille. 2015. "Career Technical Education in California's K–12 Public Schools." Unpublished manuscript.

Etzkowitz, Henry. 2003. *MIT and the Rise of Entrepreneurial Science.* London: Routledge.

Fain, Paul. 2014. "Badging from Within." *Education Week,* January 3. Accessed December 14, 2016. https://www.insidehighered.com/news/2014/01/03/uc-daviss -groundbreaking-digital-badge-system-new-sustainable-agriculture-program.

———. 2015a. "Judge Weighs In on CCSF." *Inside Higher Ed*, January 19. Accessed December 14, 2016. https://www.insidehighered.com/news/2015/01/19/san-franciscos-two-year-college-appears-less-likely-shut-down-after-court-ruling.

———. 2015b. "Trouble for an Accreditor." *Inside Higher Ed*, August 31. Accessed December 14, 2016. https://www.insidehighered.com/news/2015/08/31/californias-community-colleges-may-seek-new-accreditor.

Fayolle, Alain, and Dana T. Redford. 2014. *Handbook on the Entrepreneurial University*. Northampton, MA: Edward Elgar.

Finney, Joni E., Christina Riso, Kata Orosz, and William Casey Boland. 2014. *From Master Plan to Mediocrity: Higher Education Performance and Policy in California*. Philadelphia: Institute for Research on Higher Education.

Fischer, Claude, and Michael Hout. 2006. *Century of Difference: How America Changed in the Last One Hundred Years*. New York: Russell Sage Foundation.

Fizz, Robyn. 2012. "Open Education on the Move: An Interview with Vijay Kumar." *MIT News* (September). Accessed December 14, 2016. http://news.mit.edu/2012/open-education-on-the-move-an-interview-with-vijay-kumar-0920.

Fligstein, Neil, and Doug McAdam. 2012. *A Theory of Fields*. Oxford: Oxford University Press.

Florida, Richard, and Martin Kenney. 2000. "Transfer and Replication of Organizational Capabilities." In *The Nature and Dynamics of Organizational Capabilities*, edited by G. Dosi, R. R. Nelson, and S. G. Winter, 281–310. New York: Oxford University Press.

Foothill College. 2015. "Online Degrees and Certificates." Accessed December 14, 2016. http://www.foothill.edu/fga/degrees.php.

Foss, Lene, and David V. Gibson. 2015. *The Entrepreneurial University: Context and Institutional Change*. London: Routledge.

Freeman, Chris. 1982. *The Economics of Industrial Innovation*. London: Pinter.

Fresno Business Council. 2013. "History." Accessed December 14, 2016. http://www.fresnobc.org/about/history.

Friedland, Roger, and Robert Alford. 1991. "Bringing Society Back In: Symbols, Practices, and Institutional Contradictions." In *In the New Institutionalism in Organizational Analysis*, edited by W. W. Powell and P. J. DiMaggio, 232–63. Chicago: University of Chicago Press.

Fuller, C. 2011. *The History and Origins of Survey Items for the Integrated Postsecondary Education Data System (NPEC 2012-833)*. Washington, DC. Accessed December 14, 2016. http://nces.ed.gov/pubs2012/2012833.pdf.

Gallivan, Michael J., Duane P. Truex III, and Lynette Kvasny. 2004. "Changing Patterns in IT Skill Sets: A Content Analysis of Classified Advertising." *Data Base for Advances in Information Systems* 35 (3): 64–87. doi: 10.1145/1017114.1017121.

Geiger, Roger L. 2011. "The Ten Generations of American Higher Education." In *American Higher Education in the Twenty-First Century*, edited by P. G. Altbach, 37–68. Baltimore: Johns Hopkins University Press.

Geiser Saul, and Richard C. Atkinson. 2010. *The Case for Restructuring Baccalaureate Education in California*. Accessed December 14, 2016. http://www.cshe.berkeley.edu/beyond-master-plan-case-restructuring-baccalaureate-education-california.

Gentemann, Karen M., and James J. Fletcher. 1994. "Refocusing the Academic Program Review on Student Learning: The Role of Assessment." *New Directions for Institutional Research* 84:31–46.

Gerstein, Jackie. 2013. "I Don't Get Digital Badges." *User Generated Education.* Accessed December 14, 2016. https://usergeneratededucation.wordpress.com/2013/03/16/i -dont-get-digital-badges.

Gibbons, James F. 2000. "The Role of Stanford University: A Dean's Reflections." In *The Silicon Valley Edge: A Habitat for Innovation and Entrepreneurship,* edited by C-M. Lee, W. F. Miller, M. G. Hancock, and H. S. Rowen, 200–217. Stanford, CA: Stanford University Press.

Gilmore, C. Stewart. 2004. *Fred Terman at Stanford: Building a Discipline, a University, and Silicon Valley.* Stanford, CA: Stanford University Press.

Gillespie, Patrick. 2015. "University of Phoenix Has Lost Half Its Students." *CNN Money,* March 25. Accessed December 14, 2016. http://money.cnn.com/2015/03/25 /investing/university-of-phoenix-apollo-earnings-tank.

Giloth, Robert. 2010. *Workforce Intermediaries for the Twenty-First Century.* Philadelphia: Temple University Press.

Ginder, Scott A., Janice E. Kelly-Reid, and Farrah B. Mann. 2014. *2013–14 Integrated Postsecondary Education Data System (IPEDS) Methodology Report.* Washington, DC: National Center for Education Statistics, Institute of Education Sciences, US Department of Education. Accessed December 14, 2016. http://nces.ed.gov/pubs2014 /2014067.pdf.

Glidden, Robert. 1997. "Testimony to the Subcommittee on Postsecondary Education, Training and Lifelong Learning." *Council for Higher Education Accreditation.* Accessed December 14, 2016. http://www.chea.org/Government/Testimony/97July.asp.

Goffman, Erving. 1961. *Asylums.* Garden City, NY: Doubleday, Anchor Books.

Goodwin, Karen F., and Robert O. Riggs. 1997. "The State Postsecondary Review Program: Implications for the Community College." *Community College Journal of Research and Practice* 21 (8): 729–39. doi: 10.1080/1066892970210805.

Gordon, Howard R. D. 1999. *The History and Growth of Vocational Education in America.* Old Tappan, NJ: Prentice-Hall.

Green, Adrienne. 2015. "Will Corinthian Colleges Be Able to Pay Back Students?" *Atlantic* (October). Accessed December 14, 2016. http://www.theatlantic.com /education/archive/2015/10/corinthian-colleges-pay-back-students/413227.

Greenwood, Royston, and C. R. Hinings. 1993. "Understanding Strategic Change: The Contribution of Archetypes." *Academy of Management Journal* 36 (5): 1052–81. Accessed December 14, 2016. http://www.jstor.org/stable/256645.

Grubb, W. Norton. 2001. "From Isolation to Integration: Postsecondary Vocational Education and Emerging Systems of Workforce Development." *New Directions for Community Colleges: The New Vocationalism in Community Colleges* 115:27–37. doi: 10.1002/cc.28.

Gumport, Patricia J. 2000. "Academic Restructuring: Organizational Change and Institutional Imperatives," *Higher Education* 29:67–91.

Gutmacher, Glenn. 2000. "Secrets of Online Recruiters Exposed." *Workforce* 79 (10): 44–50. Accessed December 14, 2016. https://www.questia.com/magazine/1P3 -62300941/secrets-of-online-recruiters-exposed.

Handcock, Russell, Chris DiGiorgio, and Hon. Chuck Reed. 2013. *Index of Silicon Valley.* Joint Venture Silicon Valley Network and Silicon Valley Community Foundation. Accessed December 14, 2016. http://www.siliconvalleycf.org/sites/default/files/2013-jv -index.pdf.

Hannan, Michael T., and John H. Freeman. 1989. *Organizational Ecology.* Cambridge, MA: Harvard University Press.

Hansen, Ronald J. 2015. "Declining Enrollment at University of Phoenix Suggests Much Leaner Apollo Ahead." *Azcentral*, October 5. Accessed December 14, 2016. http://www.azcentral.com/story/money/business/2015/10/03/university-of-phoenix -enrollment-apollo-education-group-fortunes-wither/73217858.

Harkin, Tom. 2012. *For Profit Higher Education: The Failure to Safeguard the Federal Investment and Ensure Student Success.* Accessed December 14, 2016. http://www.help .senate.gov/imo/media/for_profit_report/Contents.pdf.

Harris, Brice W. 2015. *Task Force on Accreditation.* California Community Colleges Chancellor's Office. Sacramento. Accessed December 14, 2016. http:// californiacommunitycolleges.cccco.edu/Portals/0/reports/2015-Accreditation -Report-ADA.pdf.

Harris, Jeanne G., and Iris Junglas. 2013. *Decoding the Contradictory Culture of Silicon Valley.* Accenture Institute for High Performance. Accessed December 14, 2016. https://www.accenture.com/_acnmedia/Accenture/Conversion-Assets/DotCom /Documents/Global/PDF/Technology_10/Accenture-Decoding-Contradictory -Culture-Silicon-Valley.pdf.

Hasegawa, Sam. 1992. *Engineering the Future: A History of the San Jose State University College of Engineering, 1946–1991.* Sausalito, CA: Oral History Associates.

Hasse, Raimund, and George Krücken. 2013. "Competition and Actorhood: A Further Expansion of the Neo-Institutional Agenda." *Sociologia Internationalis* 51 (2): 181–205. doi: 10.3790/sint.51.2.181.

Hawley, Amos H. 1950. *Human Ecology: A Theory of Community Structure.* New York: Ronald Press.

Hellmann, Thomas. 2000. "Venture Capitalists: The Coaches of Silicon Valley." In *The Silicon Valley Edge: A Habitat for Innovation and Entrepreneurship*, edited by C-M. Lee, W. F. Miller, M. G. Hancock, and H. S. Rowen, 276–94. Stanford, CA: Stanford University Press.

Helper, Susan, Timothy Krueger, and Howard Wial. 2012. *Locating American Manufac- turing: Trends in the Geography of Production.* Washington, DC: Brookings Institution. Accessed December 14, 2016. http://www.brookings.edu/~/media/research/files /reports/2012/5/09-locating-american-manufacturing-wialh/0509_locating _american_manufacturing_report.pdf.

Henton, Doug. 2000. "A Profile of the Valley's Evolving Structure." In *The Silicon Valley Edge: A Habitat for Innovation and Entrepreneurship*, edited by C-M. Lee, W. F. Miller, M. G. Hancock, and H. S. Rowen, 46–58. Stanford, CA: Stanford University Press.

Henton, Doug, Janine Kaiser, and Kim Held. 2015. *Silicon Valley Competitiveness and Innovation Project—2015.* San Jose, CA: Silicon Valley Leadership Group and Silicon Valley Community Foundation; COECON. Accessed December 14, 2016. http:// graphics8.nytimes.com/packages/pdf/technology/SVCIP_2015_PDFfinal.pdf.

Hentschke, Guilbert C., Vicente M. Lechuga, and William G. Tierney. 2010. *For-Profit Colleges and Universities: Their Markets, Regulation, Performance, and Place in Higher Education.* Sterling, VA: Stylus.

Hillman, Nicholas W. 2014. "Differential Impacts of College Ratings: The Case of Education Deserts." Unpublished paper, University of Wisconsin-Madison, August 27. Accessed December 14, 2016. https://news.education.wisc.edu/docs /WebDispenser/news-connections-pdf/crp---hillman---draft.pdf?sfvrsn=6.

Hoffman, Reid, Ben Casnocha, and Chris Yeh. 2014. *The Alliance: Managing Talent in the Networked Age.* Boston: Harvard Business Review Press.

Holzer, Harry. 2015. *Job Market Polarization and U.S. Worker Skills: A Tale of Two Middles.* Washington, DC: Brookings Institution. Accessed December 14, 2016. http://www.brookings.edu/~/media/research/files/papers/2015/04/workforce-policy -briefs-holzer/polarization_jobs_policy_holzer.pdf.

Horn, Michael B., and Andrew P. Kelly. 2015. "Moving Beyond College: Rethinking Higher Education Regulation for an Unbundled World." August. Washington, DC: American Enterprise Institute. Accessed December 14, 2016. http://www.aei.org/wp -content/uploads/2015/08/Moving-Beyond-College.pdf.

Hossler, Don et al. 2012. *Transfer and Mobility: A National View of Pre-Degree Student Movement in Postsecondary Institutions.* National Student Clearinghouse Research Center. Herndon, VA. Accessed December 14, 2016. http://pas.indiana.edu/pdf /transfer & mobility.pdf.

Hout, Michael. 2012. "Social and Economic Returns to College Education in the United States." *Annual Review of Sociology* 38:379–400. doi: 10.1146/annurev. soc.012809.102503.

Howell, Jessica S., Michal Kurlaender, and Eric Grodsky. 2010. "Postsecondary Preparation and Remediation: Examining the Effect of the Early Assessment Program at California State University." *Journal of Policy Analysis and Management* 29 (4): 726–48. doi: 10.1002/pam.20526.

HR Focus. 2000. "Online Recruiting: What Works, What Doesn't." *HR Focus* (March): 11–13.

Hyde, Alan. 2003. *Working in Silicon Valley: Economic and Legal Analysis of a Higher Velocity Labor Market.* Armonk, NY: M. E. Sharpe.

IHS Global Insight. 2014. "U.S. Metro Economies: Income and Wage Gaps across the US." Accessed December 14, 2016. http://usmayors.org/metroeconomies/2014/08 /report.pdf.

Illich, Ivan. 1972. *Deschooling Society.* New York: Harper and Row.

Immerwahr, John. 1999a. "Doing Comparatively Well: Why the Public Loves Higher Education and Criticizes K–12." Perspectives in Public Policy: Connecting Higher Education and the Public Schools. October. Accessed December 14, 2016. http://files .eric.ed.gov/fulltext/ED437888.pdf.

———. 1999b. "Taking Responsibility: Leaders' Expectations of Higher Education." The National Center for Public Policy and Higher Education and Public Agenda. Accessed December 14, 2016. http://www.highereducation.org/reports/responsibility /responsibility.shtml.

———. 2004. "Public Attitudes on Higher Education: A Trend Analysis, 1993 to 2003." The National Center for Public Policy and Higher Education, and Public Agenda. http://www.highereducation.org/reports/pubatt.

Immerwahr, John, and Tony Foleno. 2000. *Great Expectations: How the Public and Parents— White, African American, and Hispanic—View Higher Education.* National Center for Public Policy and Higher Education, Public Agenda, Consortium for Policy Research in Education, and National Center for Postsecondary Improvement. Accessed December 14, 2016. http://www.highereducation.org/reports/expectations/expectations.shtml.

Immerwahr, John, and Jean Johnson. 2007. "Squeeze Play: How Parents And The Public Look at Higher Education Today." Accessed December 14, 2016. http://www .highereducation.org/reports/squeeze_play/squeeze_play.pdf.

Jackson, Jacob. 2014. "Higher Education in California: Student Costs." Public Policy Institute of California. Accessed December 14, 2016. http://www.ppic.org/main /publication_quick.asp?i=1121.

Jaquette, Ozan, and Edna E. Parra. 2014. "Using IPEDS for Panel Analyses: Core Concepts, Data Challenges, and Empirical Applications." In *Higher Education: Handbook of Theory and Research*, edited by M. B. Paulsen, 467–533. New York: Springer.

Jez, Su Jin. 2012. "The Role of For-Profit Colleges in Increasing Postsecondary Completions." *California Journal of Politics and Policy* 4 (2): 140–60. Accessed December 14, 2016. http://escholarship.org/uc/item/8rr5j9w4.

———. 2014. *What Data Exist That Might Be Useful to Do Research on For Profits?* Los Angeles. Accessed December 14, 2016. http://www.uscrossier.org/pullias/wp-content/uploads/2014/06/Developing-a-Research-Agenda-ALL1.pdf.

Johnson, Hans. 2010. *Higher Education in California: New Goals for the Master Plan.* Public Policy Institute of California. Accessed December 14, 2016. http://www.ppic.org/content/pubs/report/R_410HJR.pdf.

Johnson, Hans, and Marisol Cuellar Mejia. 2014. "Online Learning and Student Outcomes in California's Community Colleges." Accessed December 14, 2016. http://www.ppic.org/content/pubs/report/R_514HJR.pdf.

Johnson, Hans, Marisol Cuellar Mejia, and Sara Bohn. 2016. "Will California Run Out of College Graduates?" Public Policy Institute of California. Accessed April 8, 2017. http://www.ppic.org/content/pubs/report/R_1015HJR.pdf.

Johnson, Hans, Marisol Cuellar Mejia, and Kevin Cook. 2015. *Successful Online Courses in California's Community Colleges.* Public Policy Institute of California. Accessed December 14, 2016. http://www.ppic.org/content/pubs/report/R_615HJR.pdf.

Josh, Mitchell. 2016. "Thousands Apply to U.S. to Forgive Their Student Loans, Saying Schools Defrauded Them." *Wall Street Journal*, January 20. http://www.wsj.com/articles/thousands-apply-to-u-s-to-forgive-their-student-loans-saying-schools-defrauded-them-1453285800.

Julius, Daniel J., and Patricia J. Gumport. 2003. "Graduate Student Unionization: Catalysts and Consequences." *Review of Higher Education* 26 (2): 187–216. doi: 10.1353/rhe.2002.0033.

Kalleberg, Arne L. 2000. "Nonstandard Employment Relations: Part-Time, Temporary and Contract Work." *Annual Review of Sociology* 26 (1): 341–65. doi: 10.1146/annurev.soc.26.1.341.

Kamenetz, Anya. 2008. *Generation Debt: Why Now Is the Worst Time to Be Young.* New York: Penguin Group.

———. 2010. *DIYU: Edupunks, Edupreneurs, and the Coming Transformation of Higher Education.* White River Junction, VT: Chelsea Green.

Karabel, Jerome. 2005. *The Chosen: The Hidden History of Admission and Exclusion at Harvard, Yale, and Princeton.* New York: Houghton Mifflin.

Kelderman, Eric. 2015. "California's Community Colleges Can't Live with Accreditor, Can't Live Without It." *Chronicle of Higher Education*, December 18. Accessed December 14, 2016. http://chronicle.com/article/California-s-Community/234656.

Kena, Grace, et al. 2014. "The Condition of Education 2014." Washington, DC: National Center for Education Statistics, Institute of Education Sciences, US Department of Education. Accessed December 14, 2016. http://nces.ed.gov/pubs2014/2014083.pdf.

Kenney, Martin, ed. 2000. *Understanding Silicon Valley: The Anatomy of an Entrepreneurial Region.* Stanford, CA: Stanford University Press.

Kenney, Martin, and David C. Mowery. 2014. "Introduction." In *Public Universities and Regional Growth: Insights from the University of California*, edited by M. Kenney and D. C. Mowery, 1–19. Stanford, CA: Stanford University Press.

Kirp, David, ed. 2004. *Shakespeare, Einstein, and the Bottom Line: The Marketing of Higher Education.* Boston: Harvard University Press.

Kirst, Michael W., and Mitchell L. Stevens, eds. 2015. *Remaking College: The Changing Ecology of Higher Education.* Stanford, CA: Stanford University Press.

Kirst, Michael W., and Andrea Venezia. 2004. *From High School to College: Improving Opportunities for Success in Postsecondary Education.* San Francisco: Jossey-Bass.

Knight, Erin, and Carla Casilli. 2012. "Mozilla Open Badges." In *Game Changers: Education and Information Technologies,* edited by D. G. Oblinger, 279–84. Denver: EDUCAUSE.

Kolowich, Steve. 2013. "How edX Plans to Earn, and Share, Revenue from Its Free Online Courses." *Chronicle of Higher Education,* February 21. Accessed December 14, 2016. http://chronicle.com/article/How-EdX-Plans-to-Earn-and/137433.

Koran, Lawrence M. 1981. "Mental Health Services." In *Healthcare Delivery in the United States,* edited by S. Jonas, 235–71. New York: Springer

Kraatz, Matthew S., Marc S. Ventresca, and Lina N. Deng. 2010. "Precarious Values and Mundane Innovations: Enrollment Management in American Liberal Arts Colleges." *Academy of Management Journal* 53 (6): 1522–45. doi: 10.5465/AMJ.2010.57319260.

Kvamme, Floyd E. 2000. "Life in Silicon Valley: A First-Hand View of the Region." In *The Silicon Valley Edge: A Habitat for Innovation and Entrepreneurship,* edited by C-M. Lee, W. F. Miller, M. G. Handcock, and H. S. Rowen, 16–39. Stanford, CA: Stanford University Press.

Labaree, David F. 1997. "Public Goods, Private Goods: The American Struggle over Educational Goals." *American Educational Research Journal* 34 (1): 39–81. doi: 10.3102/00028312034001039.

Lawrence, Paul R., and Jay W. Lorsch. 1967. *Organization and Environment: Managing Differentiation and Integration.* Boston: Graduate School of Business Administration, Harvard University.

Lawrence, Thomas B., Roy Suddaby, and Bernard Leca. 2009. *Institutional Work: Actors and Agency in Institutional Studies of Organizations.* Cambridge, MA: Cambridge University Press.

Lee, Chong-Moon, William F. Miller, Marguerite G. Hancock, and Henry S. Rowen, eds. 2000. *The Silicon Valley Edge: A Habitat for Innovation and Entrepreneurship.* Stanford, CA: Stanford University Press.

Legislative Analyst's Office. 2013. "An Analysis of New Cal Grant Eligibility Rules." Accessed December 14, 2016. http://www.lao.ca.gov/reports/2013/edu/new-cal-grant/new-cal-grant-010713.pdf.

Lenoir, Timothy, et al. 2004. *Inventing the Entrepreneurial Region: Stanford and the Co-Evolution of Silicon Valley.* Stanford, CA: Stanford University Press.

Leslie, Stuart W. 2000. "The Biggest 'Angel' of Them All: The Military and the Making of Silicon Valley." In *Understanding Silicon Valley: The Anatomy of an Entrepreneurial Region,* edited by M. Kenney, 48–67. Stanford, CA: Stanford University Press.

Lewin, Tamar. 2010a. "Once a Leader, U.S. Lags in College Degrees." *New York Times,* July 23. Accessed December 14, 2016. http://www.nytimes.com/2010/07/23/education/23college.html?partner=rss&emc=rss.

———. 2010b. "U.S. Releases Rules on For-Profit Colleges." *New York Times,* July 23. Accessed December 14, 2016. http://www.nytimes.com/2010/07/23/education/23gainful.html.

———. 2012. "Instruction for Masses Knocks Down Campus Walls." *New York Times,* March 4. Accessed December 14, 2016. http://www.nytimes.com/2012/03/05/education/moocs-large-courses-open-to-all-topple-campus-walls.html?_r=0.

———. 2014. "Web-Era Trade Schools, Feeding a Need for Code." *New York Times*, October 13. Accessed December 14, 2016. http://www.nytimes.com/2014/10/14/us /web-era-trade-schools-feeding-a-need-for-code.html?nlid=22193524.

———. 2015. "For-Profit Colleges Face a Loan Revolt by Thousands Claiming Trickery." *New York Times*, May 3. Accessed December 14, 2016. http://www.nytimes .com/2015/05/04/education/for-profit-colleges-face-a-loan-strike-by-thousands -claiming-trickery.html.

Lewis, Ted G. 2000. *Microsoft Rising . . . and Other Tales of Silicon Valley*. Piscataway, NJ: Wiley-IEEE Computer Society Press.

LinkedIn. 2015. "About Us." Accessed December 14, 2016. https://press.linkedin.com /about-linkedin.

Livingston, Tab. 1998. "History of California's AB 1725 and Its Major Provisions." Accessed December 14, 2016. http://files.eric.ed.gov/fulltext/ED425764.pdf.

Long, Bridget Terry, and Michal Kurlaender. 2008. "Do Community Colleges Provide a Viable Pathway to a Baccalaureate Degree?" Cambridge, MA. National Bureau of Economic Research. Accessed December 14, 2016. http://www.nber.org/papers/w14367.

Loss, Christopher P. 2012. *Between Citizens and the State: The Politics of American Higher Education in the 20th Century*. Princeton, NJ: Princeton University Press.

Lough, Nick. 2015. "Education Officials: Some for-Profit Schools Don't Make the Grade." *WBRC Fox 6 News*, February 9. Accessed December 14, 2016. http://www.wbrc.com /story/28063806/us-dept-of-education-some-for-profit-schools-dont-make-the-grade.

Lowen, Rebecca. 1997. *Creating the Cold War University: The Transformation of Stanford*. Berkeley: University of California Press.

Lynch, Mamie, Jennifer Engle, and Jose L. Cruz. 2010. "Subprime Opportunity: The Unfulfilled Promise of For-Profit Colleges and Universities." The Education Trust. Accessed December 14, 2016. http://edtrust.org/wp-content/uploads/2013/10 /Subprime_report_1.pdf.

MacAllum, Keith, and Karla Yoder. 2004. *The 21st-Century Community College: A Strategic Guide to Maximizing Labor Market Responsiveness*. Academy for Educational Development. Accessed December 14, 2016. https://www2.ed.gov/rschstat/research /progs/ccinits/LMRvol1.doc.

Malone, Michael S. 2002. *The Valley of Heart's Delight: A Silicon Valley Notebook 1963– 2001*. New York: John Wiley and Sons.

Malone, Thomas W., Robert Laubacher, and Michael S. Scott Morton. 2003. *Inventing the Organizations of the 21st Century*. Cambridge, MA: MIT Press.

Maney, Dave. 2012. "'Badges' Fill Credential Gaps Where Higher Education Fails." *Denver Post*, March 31. Accessed December 14, 2016. http://www.denverpost.com/ci _20295793/badges-fill-credential-gaps-where-higher-education-fails.

March, James G., and Herbert A. Simon. 1958. *Organizations*. New York: John Wiley and Sons.

Marchal, Emmanuelle, Kevin Mellet, and Geraldine Rieucau. 2007. "Job Board Toolkits: Internet Match-Making and Changes in Job Advertisements." *Human Relations* 60 (7): 1091–113. doi: 10.1177/0018726707081159.

Marcus, Jon. 2005. "CUNY Sheds Reputation as 'Tutor U.'" *National Cross Talk*. Accessed December 14, 2016. http://www.highereducation.org/crosstalk/ct0205 /news0205-cuny.shtml.

Marquis, Christopher, and Julie Battilana. 2009. "Acting Globally but Thinking Locally?: The Enduring Influence of Local Communities on Organizations." *Research in Organizational Behavior* (29): 283–302. doi: 10.1016/j.riob.2009.06.001.

Marquis, Christopher, Michael Lounsbury, and Royston Greenwood, eds. 2011. *Communities and Organizations: Research in the Sociology of Organizations.* 11th ed. Bingley, UK: Emerald Group.

Martin, John Levi. 2011. *The Explanation of Social Action.* New York: Oxford University Press.

Massaro, Rachel, and Alesandra Najera. 2014. *2014 Silicon Valley Index.* Joint Venture Silicon Valley and Silicon Valley Community Foundation. Accessed December 14, 2016. http://www.siliconvalleycf.org/sites/default/files/publications/2014-silicon -valley-index.pdf.

McCann, Clare, and Amy Laitinen. 2014. *College Blackout: How the Higher Education Lobby Fought to Keep Students in the Dark.* New America. https://www.insidehighered .com/sites/default/server_files/files/CollegeAfrican Americanout_March10_Noon. pdf. Site discontinued.

McGuinness, Aims C., Jr. 2011. "The States and Higher Education." In *Education in the Twenty-First Century: Social, Political, and Economic Challenges,* edited by P. G. Altbach, P. J. Gumport, and R. O. Berdahl, 139–69. Baltimore: Johns Hopkins University Press.

McGuire, Robert. 2014. "The Best MOOC Provider: A Review of Coursera, Udacity and Edx." *Skilled Up.* Accessed December 14, 2016. http://www.skilledup.com /articles/the-best-mooc-provider-a-review-of-coursera-udacity-and-edx.

Meister, Jeanne C. 1998. *Corporate Universities: Lessons in Building a World-Class Work Force.* New York: McGraw-Hill.

Metropolitan Transportation Commission. 2015. "Commissioners." *What Is MTC?* Accessed December 14, 2016. http://mtc.ca.gov/about-mtc/what-mtc/commissioners.

Mets, Lisa A. 1995. "Lessons Learned from Program Review Experiences." *New Directions for Institutional Research* 86:81–92. doi: 10.1002/ir.37019958608

Meyer, John W. 1977. "The Effects of Education as an Institution." *American Journal of Sociology* 83 (1): 55–77. Accessed December 14, 2016. http://www.jstor.org/stable/2777763.

Meyer, John W., and Ronald L. Jepperson. 2000. "The 'Actors' of Modern Society: The Cultural Construction of Social Agency." *Sociological Theory* 18 (1): 100–120. doi: 10.1111/0735-2751.00090

Meyer, John W., and Brian Rowan. 1977. "Institutionalized Organizations: Formal Structure as Myth and Ceremony." *American Journal of Sociology* 83 (2): 55–77. Accessed December 14, 2016. http://www.jstor.org/stable/2778293.

Meyer, John W., and W. Richard Scott. 1983. *Organizational Environments: Ritual and Rationality.* Thousand Oaks, CA: Sage.

Miles, Matthew B., and Michael Huberman. 1994. *Qualitative Data Analysis: An Expanded Sourcebook.* 2nd ed. Thousand Oaks, CA: Sage.

Miller, Ben. 2014. "The College Graduation Rate Flaw That No One's Talking About." *EdCentral* (October). Accessed December 14, 2016. http://www.edcentral.org /graduation-rate-flaw.

Mintrom, Michael. 2000. *Leveraging Local Innovation: The Case of Michigan's Charter Schools.* East Lansing: Michigan State University Press.

Moore, Colleen, Su Jez, Eric Chisholm, and Nancy Shulock. 2012. "Policy Brief: Career Opportunities: Career Technical Education and the College Completion Agenda." Sacramento: California State University. March. Accessed December 14, 2016. http://files.eric.ed.gov/fulltext/ED534075.pdf.

Moretti, Enrico. 2012. *The New Geography of Jobs.* New York: Houghton Mifflin Harcourt and Mariner Books.

————. 2013. "Real Wage Inequality." *American Economic Journal: Applied Economics* 5 (1): 65–103. doi: 10.1257/app.5.1.65.

Mowery, David C., Richard R. Nelson, Bhaven N. Sampat, and Arvids A. Ziedonis. 2004. *Ivory Tower and Industrial Innovation: University-Industry Technology Transfer Before and After the Bayh-Dole Act.* Stanford, CA: Stanford University Press.

The Mozilla Foundation and Peer 2 Peer University. 2012. "Open Badges for Lifelong Learning." Mozilla University. Accessed December 14, 2016. https://wiki.mozilla .org/File:OpenBadges-Working-Paper_092011.pdf.

Mumper, Michael, Lawrence E. Gladieux, Jacqueline E. King, and Melanie E. Corrigan. 2011. "The Federal Government and Higher Education." In *American Higher Education in the Twenty-First Century: Social, Political, and Economic Challenges*, edited by P. G. Altbach, P. J. Gumport, and R. O. Berdahl, 113–38. Baltimore: Johns Hopkins University Press.

Murphy, John D. 2013. *Mission Forsaken: The University of Phoenix Affair with Wall Street.* Cambridge, MA: Proving Ground Education.

Murphy, Katy. 2014. "Corinthian Colleges, California's Largest Career-College Company, Could Go out of Business." *San Jose Mercury News*, June 19. Accessed December 14, 2016. http://www.mercurynews.com/education/ci_25998174 /corinthian-colleges-californias-largest-career-college-company-could.

Nash, George H. 1988. *Herbert Hoover and Stanford University.* Stanford, CA: Hoover Institution, Stanford University

National Center for Collective Bargaining in Higher Education and the Professions. 2006. *Directory of Faculty Contracts and Bargaining Agents in Higher Education.* New York: Baruch College, City University of New York.

National Center for Education Statistics. 2012. *The Condition of Education.* Washington, DC: US Department of Education. Accessed December 14, 2016. http://nces.ed.gov /pubsearch/pubsinfo.asp?pubid=2012045.

————. 2013. "Career/Technical Education Statistics." Washington, DC: US Department of Education. Accessed December 14, 2016. https://nces.ed.gov/surveys/ctes.

————. 2014a. "College Navigator." Washington, DC: US Department of Education. Accessed November 10, 2014. http://nces.ed.gov/collegenavigator.

————. 2014b. "Integrated Postsecondary Education System Glossary." Washington, DC: US Department of Education. Accessed November 10, 2014. http://nces.ed.gov /ipeds/glossary.

————. 2014c. "Statutory Requirements for Reporting IPEDS Data." Washington, DC: US Department of Education. Accessed November 26, 2014. https://surveys.nces.ed .gov/ipeds/ViewContent.aspx?contentId=18.

National Governors' Association. 1986. "Time for Results: The Governors' Report on Education." Washington, DC.

National Science Board. 2010. "Science and Engineering Indicators 2010." Arlington, VA: National Science Foundation. Accessed December 14, 2016. http://www.nsf.gov /statistics/seind10/pdf/seind10.pdf.

Nevens, Michael T. 2000. "Innovation in Business Models." In *The Silicon Valley Edge*, edited by C-M. Lee, W. F. Miller, M. G. Hancock, and H. S. Rowen, 81–93. Stanford, CA: Stanford University Press.

New America Foundation. 2015. *Transforming the Higher Education Act for the 21st Century.* Accessed December 14, 2016. https://static.newamerica.org /attachments/10493-transforming-the-higher-education-act-for-the-21st-century /HEA11.2.af2ba56d03b8408eb6d00299453f3d9a.pdf.

Offenstein, Jeremy and Nancy Shulock. 2009. *Community College Student Outcomes: Limitations of the Integrated Postsecondary Education Data System (IPEDS) and Recommendations for Improvement.* Institute for Higher Education Leadership and Policy, California State University, Sacramento.

Office of the Attorney General. 2013. "Attorney General Kamala D. Harris Files Suit in Alleged For-Profit College Predatory Scheme." Accessed December 14, 2016. https://oag.ca.gov/news/press-releases/attorney-general-kamala-d-harris-files-suit -alleged-profit-college-predatory.

Olivas, Michael A., and Benjamin Baez. 2011. "The Legal Environment: The Implementation of Legal Change on Campus." In *American Higher Education in the Twenty-First Century*, edited by P. G. Altback, P. J. Gumport, and R. O. Berdahl, 170–94. Baltimore: Johns Hopkins University Press.

Oliver, Christine. 1991. "Strategic Responses to Institutional Processes." *Academy of Management Review* 16 (1): 145–79. Accessed December 14, 2016. http://www.jstor .org/stable/258610.

Olneck, Michael R. 2012. "Insurgent Credentials: A Challenge to Established Institutions of Higher Education". Unpublished paper, University of Wisconsin-Madison. Accessed December 14, 2016. https://www.hastac.org/sites/default/files/documents /insurgent_credentials__michael_olneck_2012.pdf.

O'Mahony, Siobhan, and Fabrizio Ferraro. 2007. "The Emergence of Governance in an Open Source Community." *Academy of Management Journal* 50 (5): 1079–106. doi: 10.5465/AMJ.2007.27169153.

O'Mahony, Siobhan, and Karim R. Lakhani. 2011. "Organizations in the Shadow of Communities." In *Communities and Organizations: Research in Sociology of Organizations*, edited by C. Marquis, M. Lounsbury, R. Greenwood, 3–36. Bingley, UK: Emerald Group.

O'Mara, Margaret Push. 2005. *Cities of Knowledge: Cold War Science and the Search for the Next Generation Silicon Valley.* Princeton, NJ: Princeton University Press.

The Open University. 2014. "Facts and Figures 2014/15." Accessed December 14, 2016. http://www.open.ac.uk/about/main/sites/www.open.ac.uk.about.main/files/files/fact _figures_1415_uk.pdf.

Osterman, Paul. 2010. "The Promise, Performance, and Policy of Community Colleges." In *Reinventing Higher Education: The Promise of Innovation*, edited by B. Wildavsky, A. P. Kelly, and K. Carey, 129–59. Cambridge, MA: Harvard Education Press.

Palomba, Catherine A., and Trudy W. Banta. 1999. *Assessment Essentials: Planning, Implementing, and Improving Assessment in Higher Education.* San Francisco: Jossey-Bass.

Pastor, Manuel, Rhonda Ortiz, Marlene Ramos, and Mirabai Auer. 2012. "Immigrant Integration: Integrating New Americans and Building Sustainable Communities." Equity Issue Brief. Los Angeles: University of Southern California. https://www .policylink.org/sites/default/files/immigrant_integration_brief.pdf.

Patterson, Wayne. 1999. "Certificate Programs Raise Important Issues." *CGS Communicator* (April): 1–3.

Peele, Thomas, and Chris De Benedetti. 2014. "Career College Chain to Close amid Scandal." *San Jose Mercury News*, June 24. Accessed December 14, 2016. http://www .pressreader.com/usa/san-jose-mercury-news/20140624/textview.

Peralta Community College District. 2016. "Career Technical Education Home." *Peralta Colleges.* Accessed December 14, 2016. http://web.peralta.edu/cte.

Peterson, Marvin W. 2007. "The Study of Colleges and Universities as Organizations." In *Sociology of Higher Education: Contributions and Their Contests*, edited by P. J. Gumport, 147–84. Baltimore: Johns Hopkins University Press.

Petraeus, Hollister K. 2011. "For-Profit Colleges, Vulnerable G.I.'s." *New York Times*, September 21. Accessed December 14, 2016. http://www.nytimes.com/2011/09/22/opinion/for-profit-colleges-vulnerable-gis.html.

Pfeffer, Jeffrey, and Gerald R. Salancik. 1978. *The External Control of Organizations: A Resource Dependence Perspective*. New York: Harper and Row.

Piore, Michael J., and Charles F. Sabel. 1984. *The Second Industrial Divide: Possibilities for Prosperity*. New York: Basic Books.

Porter, Eduardo. 2014. "A Smart Way to Skip College in Pursuit of a Job: Udacity-AT&T 'NanoDegree' Offers an Entry-Level Approach to College." *New York Times*, June 17. Accessed December 14, 2016. http://www.nytimes.com/2014/06/18/business/economy/udacity-att-nanodegree-offers-an-entry-level-approach-to-college.html?_r=0.

Powell, Arthur G., Eleanor Farrar, and David K. Cohen. 1985. *The Shopping Mall High School: Winners and Losers in the Educational Marketplace*. Boston: Houghton Mifflin.

Powell, Walter W. 1990. "Neither Market nor Hierarchy: Network Forms of Organization." In *Research in Organizational Behavior*, vol. 12, edited by B. M. Staw and L. Cummings, 295–336. Greenwich, CT: JAI Press.

Powell, Walter W., Kenneth W. Koput, and Laurel Smith-Doerr. 1996. "Interorganizational Collaboration and the Locus of Innovation: Networks of Learning in Biotechnology." *Administrative Science Quarterly* 41 (1): 116–45. doi: 10.2307/2393988.

Powell, Walter W., Kelley Packalen, and Kjersten Whittington. 2012. "Organizational and Institutional Genesis: The Emergence of Hi-Tech Clusters in the Life Sciences." In *The Emergence of Organizations and Markets*, edited by J. F. Padgett and W. W. Powell, 434–65. Princeton, NJ: Princeton University Press.

Powell, Walter W., and Kaisa Snellman. 2004. "The Knowledge Economy." *Annual Review of Sociology* 30:199–220. doi: 10.1146/annurev.soc.29.010202.100037.

Public Agenda. 1999. *Kids These Days '99: What Americans Really Think about the Next Generation*. Accessed December 14, 2016. http://www.publicagenda.org/files/kids_these_days_99.pdf.

Public Policy Institute of California. 2013. "Immigrants in California." Accessed December 14, 2016. http://www.ppic.org/main/publication_show.asp?i=258.

Quinn, Michelle. 2014. "The Shuttle Effect, and the Commute That Divides Us." *San Jose Mercury News*, December 5. Accessed December 14, 2016. http://www.mercurynews.com/michelle-quinn/ci_27078188/quinn-commute-that-divides-us.

Ramirez, Francisco O., and John Boli. 1987. "Global Patterns of Educational Institutionalization." In *Institutional Structure: Constituting State, Society, and the Individual*, edited by G. M. Thomas, J. W. Meyer, F. O. Ramirez, and J. Boli, 150–72. Newbury Park, CA: Sage.

Randolph, Sean. 2012. *The Bay Area Innovation System*. San Francisco: Bay Area Council Economic Institute. Accessed December 14, 2016. http://www.bayeconfor.org/media/files/pdf/BayAreaInnovationSystemWeb.pdf.

Randolph, Sean, and Hans Johnson. 2014. *Reforming California Public Higher Education for the 21st Century*. A Bay Area Council Economic Institute White Paper. San Francisco. Accessed December 14, 2016. http://www.bayareaeconomy.org/report/reforming-california-public-higher-education-for-the-21st-century.

Resnick, Mitchel. 2012. "Still a Badge Skeptic." *Hastac*, February 27. Accessed December 14, 2016. https://www.hastac.org/blogs/mres/2012/02/27/still-badge-skeptic.

Richardson, Richard C., Kathy Reeves-Bracco, Patrick M. Callan, and Joni E. Finney. 1999. *Designing State Higher Education Systems for a New Century.* Phoenix, AZ: American Council on Education/Oryx Press.

Richardson, Richard Jr., and Mario Martinez. 2009. *Policy and Performance in American Higher Education: An Examination of Cases across State Systems.* Baltimore: Johns Hopkins University Press.

Rockhill, Kathleen. 1983. *Academic Excellence versus Public Service: The Development of Adult Higher Education in California.* New Brunswick, NJ: Transaction Books.

Roland, Gerard. 2004. "Understanding Institutional Change: Fast-Moving and Slow-Moving Institutions." *Studies in Comparative International Development* 38 (4): 109–31.doi: 10.1007/BF02686330.

Rosenbaum, James E., Regina Deil-Amen, and Ann E. Person. 2006. *After Admission: From College Access to College Success.* New York: Russell Sage Foundation.

Rosenkopf, Lori, and Michael L. Tushman. 1998. "The Coevolution of Community Networks and Technology: Lessons from the Flight Simulation Industry." *Industrial and Corporate Change* 7 (2): 311–46. doi: 10.1093/icc/7.2.311.

Ruef, Martin, and Manish Nag. 2015. "The Classification of Organizational Forms." In *Remaking College: The Changing Ecology of Higher Education*, edited by M. W. Kirst and M. L. Stevens, 84–109. Stanford, CA: Stanford University Press.

Ruiz, Neil G. 2014. *The Geography of Foreign Students in U.S. Higher Education: Origins and Destinations.* Global Cities Initiative. Washington, DC: Brookings Institution. http://www.brookings.edu/~/media/research/files/reports/2014/08/foreign students /foreign_students_final.pdf. Site discontinued.

Ruiz, Neil G., Jill H. Wilson, and Shyamali Choudhury. 2012. *The Search for Skills: Demand for H-1B Immigrant Workers in US Metropolitan Areas.* Washington, DC: Brookings Institution. Accessed December 14, 2016. http://immigrationresearch -info.org/system/files/Brookings---Search_for_Skills_H1B_Visas.pdf.

Salzman, Hal, Daniel Kuehn, and B. Lindsay Lowell. 2013. "Guestworkers in the High-Skill U.S. Labor Market: An Analysis of Supply, Employment and Wage Trends." *Economic Policy Institute* 359:1–35. Accessed December 14, 2016. http://dx.doi .org/doi:10.7282/T379469D.

San Jose Mercury News. 2013. "How For-Profit Colleges in the Bay Area Measure Up." 1–2. http://www.mercurynews.com/news/ci_24041129/how-profit-colleges-bay-area -measure-up.

San José State University. 2016. "Points of Pride." *San José State University.* Accessed December 14, 2016. http://www.sjsu.edu/about_sjsu/pride.

Saxenian, AnnaLee. 1996. *Regional Advantage: Culture and Competition in Silicon Valley and Route 128.* Cambridge, MA: Harvard University Press.

———. 2000a. "Networks of Immigrant Entrepreneurs." In *The Silicon Valley Edge*, edited by C-M. Lee, W. F. Miller, M. G. Handcock, H. S. Rowen, 248–68. Stanford, CA: Stanford University Press.

———. 2000b. "The Origins and Dynamics of Production Networks in Silicon Valley." In *Understanding Silicon Valley: Anatomy of an Entrepreneurial Region*, edited by Martin Kenney, 141–62. Stanford: Stanford University Press.

———. 2000c. "The Role of Immigrant Entrepreneurs in New Venture Creation." In *The Entrepreneurship Dynamic: Origins of Entrepreneurship and the Evolution of Industries*, edited by C. B. Schoonhoven and E. Romanelli, 68–108. Stanford, CA: Stanford University Press.

———. 2002. "Transnational Communities and the Evolution of Global Production Networks: The Cases of Taiwan, China, and India." *Industry and Innovation* 9 (3): 183–202

———. 2006. *The New Argonauts: Regional Advantage in a Global Economy*. Cambridge, MA: Harvard University Press.

———. 2008. "International Mobility of Engineers and the Rise of Entrepreneurship in the Periphery." In *The International Mobility of Talent: Types, Causes, and Development Impact*, edited by Andrés Solimano, 117–44. London: Oxford University Press.

Saxenian, AnnaLee, Yasuyuki Motoyama, and Xiaohong Quan. 2002. *Local and Global Networks of Immigrant Professionals in Silicon Valley*. San Francisco: Public Policy Institute of California.

Schramm, Wilbur. 1962. "What We Know about Learning from Instructional Television." In *Educational Television: The Next Ten Years*, edited by L. Asheim et al., 52–76. Stanford, CA: Institute for Communication Research.

Scott, W. Richard. 1977. "Effectiveness of Organizational Effectiveness Studies." In *New Perspectives on Organizational Effectiveness*, edited by P. S. Goodman and J. M. Pennings, 63–95. San Francisco: Jossey-Bass.

———. 1985. "Systems within Systems; The Mental Health Sector." In *The Organization of Mental Health Services*, edited by W. Richard Scott and Bruce L. Black, 31–52. Beverly Hills: Sage.

———1992. "The Organization of Environments: Network, Cultural, and Historical Elements." In *Organization Environments: Ritual and Rationality*, updated edition, edited by John W. Meyer and W. Richard Scott, 155–75. Newbury Park, CA: Sage.

———. 2008. "Lords of the Dance: Professionals as Institutional Agents." *Organization Studies* 29 (2): 219–38. doi: 10.1177/0170840607088151.

———. 2014. *Institutions and Organizations: Ideas, Interests, and Identities*. 4th ed. Los Angeles: Sage.

———. 2015. "Higher Education in America: Multiple Field Perspectives." In *Remaking College: The Changing Ecology of Higher Education*, edited by M.W. Kirst and M. L. Stevens, 19–38. Stanford, CA: Stanford University Press.

Scott, W. Richard, and Manuelito Biag. 2016. "The Changing Ecology of Higher Education: An Organizational Field Perspective." In *The University under Pressure: Research in the Sociology of Organizations*, vol. 46, edited by E. P. Berman and C. Paradeise, 25–51. Bingley, UK: Emerald Group.

Scott, W. Richard, and Gerald F. Davis. 2007. *Organizations and Organizing: Rational, Natural, and Open System Perspectives*. Upper Saddle River, NJ: Pearson Prentice Hall.

Scott, W. Richard, and John W. Meyer. 1992. "The Organization of Societal Sectors." In *Organizational Environments: Ritual and Rationality*, edited by J. W. Meyer and W. R. Scott, 129–54. Newbury Park, CA: Sage.

Seely-Brown, John, and Paul Duguid. 1991. "Organizational Learning and Communities-of-Practice: Toward a Unified View of Working, Learning, and Innovation." *Organizational Science* 2 (1): 40–57. Accessed December 14, 2016. http://www.jstor.org/stable/2634938.

———. 2000. "Mysteries of the Region: Knowledge Dynamics in Silicon Valley." In *The Silicon Valley Edge: A Habitat for Innovation and Entrepreneurship*, edited by C-M. Lee, W. F. Miller, M. G. Hancock, and H. S. Rowen, 16–39. Stanford, CA: Stanford University Press.

Seifert, Tricia E., Ernest T. Pascarella, Sherri I. Erkel, and Kathleen M. Goodman. 2011. "The Importance of Longitudinal Pretest-Posttest Designs in Estimating

College Impact." In *Longitudinal Assessment for Institutional Improvement: New Directions for Institutional Research, Assessment Supplement 2010*, edited by T. Seifert, 5–16. San Francisco: Jossey-Bass.

Selingo, Jeffrey J. 2015. "Finding a Career Track in LinkedIn Profiles." *New York Times*, July 31. Accessed December 14, 2016. http://www.nytimes.com/2015/08/02/education/edlife/finding-direction-in-linkedin-profiles.html?_r=0.

Settersten, Richard R., Jr. 2015. "The New Landscape of Early Adulthood: Implications for Broad-Access Higher Education." In *Remaking College: The Changing Ecology of Higher Education*, edited by Michael W. Kirst and Mitchell L. Stevens, 113–33. Stanford, CA: Stanford University Press.

Sewell, William H., and Robert M. Hauser. 1975. *Education, Occupation, and Earnings*. New York: Academic Press.

Shaw, Robert. 1993. "A Backward Glance: To a Time Before There Was Accreditation." *NCA Quarterly* 68 (2): 323–35. Accessed December 14, 2016. http://eric.ed.gov/?id=EJ483834.

Shulock, Nancy, Jodi Lewis, and Connie Tan. 2013. "Workforce Investments: State Strategies to Preserve Higher-Cost Career Education Programs in Community and Technical Colleges." Institute for Higher Education Leadership and Policy. August. Accessed December 14, 2016. http://doingwhatmatters.cccco.edu/portals/6/docs/IHELP_Workforce_Invest_FINAL_Aug30.pdf.

Shulock, Nancy, and Colleen Moore. 2013. "Career Opportunities: Career Technical Education and the College Completion Agenda. Part IV: Aligning Policy with Mission for Better Outcomes. Institute for Higher Education Leadership & Policy." Institute for Higher Education Leadership and Policy. March. Accessed December 14, 2016. http://edinsightscenter.org/Portals/0/ReportPDFs/career-opportunities-part-4.pdf.

Shulock, Nancy, Colleen Moore, and Jeremy Offenstein. 2011. *The Road Less Traveled: Realizing the Potential of Career Technical Education in the California Community Colleges*. Institute for Higher Education Leadership and Policy. Accessed December 14, 2016. http://files.eric.ed.gov/fulltext/ED524217.pdf.

Shulock, Nancy, Colleen Moore, and Connie Tan. 2014. "A New Vision for California Higher Education: A Model Public Agenda Executive Summary." Sacramento State Institute for Higher Education Leadership and Policy. Accessed December 14, 2016. http://edinsightscenter.org/Portals/0/ReportPDFs/a-new-vision-for-california-highered.pdf?ver=2016-01-15-155401-863.

Siegfried, John J., Allen R. Sanderson, and Peter McHenry. 2007. "The Economic Impact of Colleges and Universities." *Economics of Education Review* 26 (5): 546–58. doi: 10.3200/CHNG.40.2.24-31.

Silicon Valley Institute for Regional Studies. 2015. "Income Inequality in the San Francisco Bay Area." Accessed December 14, 2016. http://siliconvalleyindicators.org/pdf/income-inequality-2015-06.pdf.

Simon, Herbert A. 1945/1977. *Administrative Behavior: A Study of Decision-Making Processes in Administration Organizations*. 4th ed. New York: Free Press.

Slaughter, Sheila, and Larry L. Leslie. 1997. *Academic Capitalism: Politics, Policies and the Entrepreneurial University*. Baltimore: Johns Hopkins University Press.

Smith, Ashley A. 2015. "Reshaping the For-Profit." *Inside Higher Ed*, July 15. Accessed December 14, 2016. https://www.insidehighered.com/news/2015/07/15/profit-industry-struggling-has-not-reached-end-road.

Smith, Burck. 2013. "Keeping College within Reach: Improving Higher Education through Innovation." Washington, DC: House Education and the Workforce Committee. Accessed December 14, 2016. http://edworkforce.house.gov/uploadedfiles/smith_testimony_final.pdf.

Snyder, Thomas D., and Sally A. Dillow. 2012. *Digest of Educational Statistics 2011*. Washington, DC. Accessed December 14, 2016. http://nces.ed.gov/pubs2012/2012001_0.pdf.

Spellings, Margaret. 2006. "A Test of Leadership. Charting the Future of U.S. Higher Education." US Department of Education. Accessed December 14, 2016. http://www2.ed.gov/about/bdscomm/list/hiedfuture/reports/final-report.pdf.

Sperling, John. 2000. *Rebel with a Cause: The Entrepreneur Who Created the University of Phoenix and the For-Profit Revolution in Higher Education*. New York: John Wiley and Sons.

Stadtman, Verne A. 1967. *The Centennial Record of the University of California*. Berkeley, CA: Centennial.

Stark, David. 1996. "Recombinant Property in East European Capitalism." *American Journal of Sociology* 101 (4): 993–1027. doi: 10.1086/230786.

Steinacker, Annette. 2005. "The Economic Effect of Urban Colleges on Their Surrounding Communities." *Urban Studies* 42 (7): 1161–75. doi: 10.1080/00420980500121335.

Stevens, Mitchell L. 2015. "Introduction: The Changing Ecology of U.S. Higher Education." In *Remaking College: The Changing Ecology of Higher Education*, edited by M. W. Kirst and M. L. Stevens, 1–18. Stanford, CA: Stanford University Press.

Stevens, Mitchell L., Elizabeth A. Armstrong, and Richard Arum. 2008. "Sieve, Incubator, Temple, Hub: Empirical and Theoretical Advances in the Sociology of Higher Education." *Annual Review of Sociology* 34 (1): 127–51. doi: 10.1146/annurev.soc.34.040507.134737.

Stocking, Carol. 1985. "The United States." In *The School and the University: An International Perspective*, edited by Burton R. Clark, 261. Berkeley: University of California Press.

Stone, Deborah. 2002. *Policy Paradox: The Art of Political Decision Making*. 3rd ed. New York: W. W. Norton.

Sturgeon, Timothy. 2000. "How Silicon Valley Came to Be." In *Understanding Silicon Valley: Anatomy of an Entrepreneurial Region*, edited by M. Kenney, 15–47. Stanford, CA: Stanford University Press.

Suchman, Mark C. 1995. "Managing Legitimacy: Strategic and Institutional Approaches." *Academy of Management Review* (20): 571–610. doi: 10.5465/AMR.1995.9508080331.

———. 2000. "Dealmakers and Counselors: Law Firms as Intermediaries in the Development of the Silicon Valley." In *Understanding Silicon Valley: The Anatomy of an Entrepreneurial Region*, edited by M. Kenney, 71–98. Stanford, CA: Stanford University Press.

Thelin, John R., and Marybeth Gasman. 2010. "Historical Overview of American Higher Education." In *Student Services: A Handbook for the Profession*, edited by S. R. Harper, 3–22. San Francisco: Jossey-Bass.

Thille, Candace, John Mitchell, and Mitchell Stevens. 2015. "What We've Learned from MOOCs." *Inside Higher Ed*, September 22. Accessed December 14, 2016. https://www.insidehighered.com/views/2015/09/22/moocs-are-no-panacea-they-can-help-improve-learning-essay.

Thompson, James D. 1967. *Organizations in Action: Social Science Bases of Administrative Theory*. New York: McGraw-Hill.

Thornton, Patricia H., and William Ocasio. 2008. "Institutional Logics." In *The Sage Handbook of Organizational Institutionalism*, edited by R. Greenwood, C. Oliver, K. Sahlin, and R. Suddaby, 99–12. Los Angeles: Sage.

Thornton, Patricia H., William Ocasio, and Michael Lounsbury. 2012. *The Institutional Logics Perspective: A New Approach to Culture, Structure and Process*. New York: Oxford University Press.

Tierney, Willam G., and Guilbert C. Hentschke. 2007. *New Players, Different Game: Understanding the Rise of For-Profit Colleges and Universities*. Baltimore: Johns Hopkins University Press.

Tinto, Vincent. 2004. "Linking Learning and Leaving." In *Reworking the Student Departure Puzzle*, edited by J. M. Braxton, 81–94. Nashville: Vanderbilt University Press.

Turner, Sarah E. 2006. "For-Profit Colleges in the Context of the Market for Higher Education." In *Earnings from Learning: The Rise of For-Profit Universities*, edited by D. W. Breneman, B. Pusser, and S. E. Turner, 54–55. Albany: SUNY Press.

Tushman, Michael L., and Charles O'Reilly. 2011. "Organizational Ambidexterity in Action: How Managers Explore and Exploit." *California Management Review* 53 (4): 5–22.

Tyack, David, and Larry Cuban. 1997. *Tinkering toward Utopia: A Century of Public School Reform*. Boston: Harvard University Press.

US Department of Education. 2015. "Information on Debt Relief for Students at Corinthian Colleges (Everest, Heald, and WyoTech)." Federal Student Aid. Accessed December 14, 2016. https://studentaid.ed.gov/sa/about/announcements/corinthian.

US Department of Education. 2016. "The Database of Accredited Postsecondary Institutions and Programs." Office of Postsecondary Education. Accessed December 14, 2016. ope.ed.gov/accreditation.

US Department of Labor. Bureau of Labor Statistics. 2015. "Labor Force Characteristics of Foreign-Born Workers Summary." Accessed December 14, 2016. http://www.bls.gov/news.release/forbrn.nro.htm.

———. 2012. "Occupational Employment Statistics." Accessed December 14, 2016. http://www.bls.gov/oes/2012/may/oes_stru.htm.

Van de Ven, Andrew H. 2005. "Running in Packs to Develop Knowledge-Intensive Technologies." *MIS Quarterly* 29 (2): 365–78. Accessed December 14, 2016. http://www.jstor.org/stable/25148683.

VanOverbeke, Mark. 2008. *The Standardization of American Schooling: Linking Secondary and Higher Education, 1870–1910*. New York: Palgrave Macmillan.

Veysey, Laurence R. 1965. *The Emergence of the American University*. Chicago: University of Chicago Press.

Walker, C. S. 2013. "New Growth in Higher Ed." *SJSU Washington Square*. Accessed December 14, 2016. https://blogs.sjsu.edu/wsq/2013/10/20/new-growth-in-higher-ed.

Walters, Dan. 2013. "University of Phoenix's Political Tale Runs Full Circle." *Pasadena Star News*, February 21. Accessed December 14, 2016. http://www.pasadenastarnews.com/opinion/20130121/university-of-phoenixs-political-tale-runs-full-circle-opinion.

———. 2015. "Hot Bay Area Economy Props Up California." *Sacramento Bee*, September 25. Accessed December 14, 2016. sacbee.com/news/politics-government/capitol-alert/article36633297.html.

Washburn, Jennifer. 2005. *University Inc.: The Corporate Corruption of American Higher Education*. New York: Basic Books.

Washburn, Jennifer, ed. 2005. "The Lessons of History." In *University Inc.: The Corporate Corruption of American Higher Education*, 25–49. New York: Basic Books.

Wedlin, Linda. 2006. *Ranking Business Schools: Forming Fields, Identities and Boundaries in International Management Education*. Cheltenham, UK: Edward Elgar.

Weick, Karl E. 1976. "Educational Organizations as Loosely Coupled Systems." *Administrative Science Quarterly* 21 (March): 1–19. Accessed December 14, 2016. http://www.jstor.org/stable/2391875.

Weinbren, Daniel. 2014. *The Open University: A History*. New York: Palgrave Macmillan.

Weisbrod, Burton A., Jeffrey P. Ballou, and Evelyn D. Asch. 2008. *Mission and Money: Understanding the University*. Cambridge: Cambridge University Press.

Western Association of Schools and Colleges (WASC). 2015. "ACS WASC Commission." Accrediting Commission for Schools Western Association of Schools and Colleges. Accessed December 14, 2016. http://www.acswasc.org/wasc/the -commission.

White, Lisa P. 2015. "Peralta Community Colleges Face Accreditation Problems." *San Jose Mercury News*, July 10. Accessed December 14, 2016. http://www.mercurynews .com/breaking-news/ci_28466597/peralta-community-colleges-face-accreditation -problems.

Wilson, B. 2015a. "Bay Area Income Gap Now More than $250,000 between Top and Bottom." *KQED News*, June 29. Accessed December 14, 2016. http://ww2.kqed.org /news/2015/06/29/bay-area-income-gap-now-more-than-250000-between-top-and -bottom.

Wilson, B. 2015b. "Growing Labor Movement Shakes Up Silicon Valley." *KQED News*, July 14. http://ww2.kqed.org/news/2015/07/14/growing-labor-movement-shakes-up -silicon-valley.

Wilson, Harold. 1963. "Harold Wilson." *Spartacus Educational*. http://spartacus -educational.com/PRwilsonHa.htm.

Wong, Queenie. 2015. "LinkedIn and Lynda Aim to Close a Skills Gap." *San Jose Mercury News*, August 7. Accessed December 14, 2016. http://www.mercurynews .com/business/ci_28603887/linkedin-and-lynda-aim-close-skills-gap.

Young, Kenneth E., Charles M. Chambers, and Herbert R. Kells. 1983. *Understanding Accreditation*. San Francisco: Jossey-Bass.

Zakaria, Fareed. 2015. *In Defense of a Liberal Education*. New York: W. W. Norton.

Zemsky, Robert. 2009. *Making Reform Work: The Case for Transforming American Higher Education*. Piscataway, NJ: Rutgers University Press.

Zumeta, William. 2001. "Public Policy and Accountability in Higher Education: Lessons from the Past and Present for the New Millennium." In *The States and Public Higher Education Policy: Affordability, Access, and Accountability*, edited by D. E. Heller, 155–97. Baltimore: Johns Hopkins University Press.

Zumeta, William, David W. Breneman, Patrick Callan, and Joni E. Finney. 2012. *Financing American Higher Education in the Era of Globalization*. Cambridge, MA: Harvard University Press.

Contributors

W. Richard (Dick) Scott is Professor Emeritus of Sociology with courtesy appointments in the Graduate School of Business, Graduate School of Education, School of Engineering, and School of Medicine, Stanford University. He received his undergraduate and MA degree from the University of Kansas and his PhD from Chicago. He has taught at Stanford his entire career. He is the author of two widely employed texts, *Organizations and Organizing: Rational, Natural and Open Systems Perspectives* (with G. F. Davis), now in its sixth edition, and *Institutions and Organizations: Ideas, Interests, and Identities,* now in its fourth edition. He is a long-term student of professional organizations and has authored or coauthored books such as *Evaluation and the Exercise of Authority* (1975), *Hospital Structure and Performance* (1987), *Institutional Change and Health-care Organizations* (2000), and *Between Movement and Establishment: Organizations Advocating for Youth* (2009). Since the 1970s, he has contributed to the development of institutional theory as applied to organizations with books such as *Organizational Environments: Ritual and Rationality* (1992), and *Institutional Environments and Organizations: Structural Complexity and Individualism* (1994). Scott was elected to the National Academy of Medicine in 1975 and received the Distinguished Scholar Award (1988) and the Distinguished Educator Award (2013) from the Management and Organization Theory Division of the Academy of Management. In 1996, he received the Distinguished Scholarly Career Award from the Academy of Management and was named Eminent Scholar of the Year by the Academy of International Business in 2015. He has received three honorary degrees.

Michael W. Kirst is Professor Emeritus of Education and Business Administration (by courtesy) at Stanford University, and president of the California State Board of Education. He has been on the Stanford faculty since 1969. Kirst received his PhD in political economy and government from Harvard. Before joining the Stanford University faculty, Kirst held several positions with the federal government, including staff director of the US Senate Subcommittee on Manpower, Employment, and Poverty, and director of Program Planning for Elementary and Secondary Education of the US Office of Education. He was a president of the California State Board of Education from 1977 to 1981. His two latest books are *From High School to College*, coedited with Andrea Venezia (2009), and *Remaking College: The Changing Ecology of Higher Education* (2015), coedited with Mitchell L. Stevens. His book *The Political Dynamics of American Education*, coauthored with Frederick M. Wirt, has been in publication since 1972, with the latest version in 2009. Kirst is a member of the National Academy of Education and the International Academy of Education.

Manuelito Biag is an associate in improvement science at the Carnegie Foundation for the Advancement of Teaching. He seeks to bridge academia and policy by investigating the implementation and influence of reforms in K–12 and higher education on students' learning and development. His research has been presented in community forums and professional conferences and published in academic journals, policy briefs, and edited volumes. He holds a PhD in education policy from the University of California, Davis.

Brian Holzman is a postdoctoral fellow at the Houston Education Research Consortium at Rice University. Using quantitative methods, he seeks to understand racial/ethnic and socioeconomic inequalities in educational attainment and to evaluate policies and interventions that can reduce gaps between groups. His ongoing research examines in-state resident tuition policies for undocumented students, summer melt, social interactions in high school, and college knowledge. He holds a PhD in sociology of education and higher education from Stanford University.

Bernardo Lara is a faculty member at the School of Business and Economics at the University of Talca in Chile. His research focuses on the application of public finance tools to understand educational systems and inequality. He holds a PhD in the economics of education and an MA in economics from Stanford University.

Judy C. Liang is an educator at the Kang Chiao International School in Taipei, Taiwan. She teaches middle school mathematics and leads data analysis initiatives. She has conducted research examining K–12 and postsecondary policies on student achievement. She is passionate about work that improves education standards and pushes for education equity for all students. She holds an MA in policy, organization, and leadership studies from Stanford University.

Anne Podolsky is a researcher and policy analyst at the Learning Policy Institute. She focuses on improving educational opportunities and outcomes, especially for students from underserved communities. Podolsky earned her MA in education policy from Stanford, JD from the University of San Diego, and BS in elementary education from Loyola University Chicago, where she graduated summa cum laude. Podolsky is an Illinois State Board of Education–certified teacher and a member of the State Bar of California.

Ethan Ris is a PhD candidate at Stanford University's Graduate School of Education. He is a historian who studies the ways in which policy elites in government and philanthropic foundations have attempted to reshape the form and function of undergraduate education in the United States. His work has been published in the *Journal of Higher Education*, *History of Education*, and the *Journal of Educational Controversy*.

Laurel Sipes is the senior policy analyst at the John W. Gardner Center for Youth and Their Communities at Stanford University. She works to build stronger ties among the education policy, research, and practice sectors to address pressing challenges in early learning, K–12, and postsecondary education. Her work also includes building organizational capacity around strategic thinking and data use. She holds an MA in public policy from the University of California, Berkeley.

Index